BEFORE IT IS
TOO LATE

Also by Daisaku Ikeda and published by Macdonald:

HUMAN VALUES IN A CHANGING WORLD: A dialogue on the Social Role of Religion written with Bryan Wilson

BEFORE IT IS TOO LATE

A Dialogue between
Aurelio Peccei and Daisaku Ikeda

Macdonald

A Macdonald Book

Copyright © Daisaku Ikeda and Aurelio Peccei 1984

First published in Great Britain in 1984 by
Macdonald & Co (Publishers) Ltd
London & Sydney

British Library Cataloguing in Publication Data

Ikeda, Daisaku
Before it is too late.
1. Environmental protection   2. Ecology
I. Title   II. Peccei, Aurelio
304.2   TD170.2

ISBN 0-356-10478-8

Photoset in North Wales by
Derek Doyle & Associates, Mold, Clwyd
Printed in Great Britain by
Redwood Burn Limited, Trowbridge, Wiltshire
Bound at the Dorstel Press

Macdonald & Co (Publishers) Ltd
Maxwell House
74 Worship Street
London EC2A 2EN

A BPCC plc Company

CONTENTS

FOREWORD

This book is about man's condition and future prospects both as they have evolved from his past thinking, behavior, deeds and from events beyond his control, and as they can be influenced by current trends and human action from now on.

We, the authors, are laymen – one from Japan, committed to expanding the teachings and practice of Buddhism, the other from Europe, committed to improving pragmatic, secular understanding of the human *problematique*. We both believe and hope that, in our way, we can contribute – no matter how little – to the good of man in these troubled times.

We recognize the differences between our viewpoints in certain fields and the somewhat divergent interpretations we give to particular situations or exigencies. We concede that we cannot always be in complete accord on aims and means of attaining them. However, we do share a vast area of common ground and are of the opinion that in-depth exploration of our shared views will enlarge and refine our thinking. We believe, too, that open discussion based on mutual respect of the differences between people can be enlightening and enriching, and can clear new avenues for further reflection.

For these reasons, we think it useful to share some of our ideas with others! In submitting our views for criticism and debate to readers in various parts of the world, we consider it appropriate to outline certain attitudes and actions that seem conducive to the improvement of the human lot in the years and decades to come. In the face of the magnitude and complexity of the issues confronting our generations, we do this in all humility. And, instead of proposing direct answers, we prefer to suggest methods of approach and the outlook we judge most likely to lead, eventually, to appropriate solutions.

We want to stress, however, our firm conviction that an adequate human response to the threats and challenges that have arisen at this perhaps decisive point in history must not be delayed much longer. The overall world situation is getting no better, and the danger that current crises may become much graver is very real. On the other hand, it is absolutely unjustifiable both to continue neglecting the immense opportunities humankind has been given as a consequence of scientific knowledge and technical means, and to forfeit our moral obligation to strive constantly to improve our lot.

We agree that today's major problems are still spiritual and ethical and that no amount of scientific and technological power or economic means can solve them. They pertain to the innermost human spheres and, only if each human being profoundly renews his visions and values, will we be able to understand and approach them with the serene hope that, if we improve ourselves from within, we will never be overwhelmed.

The urgency of the human awakening that we feel is possible and the faith we have in the human revolution that may prompt that awakening have inspired us to give this book the title *Before It Is Too Late*.

Aurelio Peccei

Daisaku Ikeda

Part 1 MAN AND NATURE

Views of Aurelio Peccei

MAN AND NATURE

Views of Aurelio Peccei

Although age-old questions inherited from eras forgotten in the depths of the past still haunt us, even more puzzling than these old ghosts are the new clusters of problems that are continually emerging and intertwining with the old ones in our time. If world situations were already baffling a couple of decades ago, they are now totally disconcerting and laden with ever-new challenges. In spite of the enormous amount of knowledge and technical means at our disposal, we are beset by a feeling of inadequacy that erodes the confidence we recently acquired in our own power and resourcefulness, and even creates a sense of insuperable crisis. In order to assess both how serious our present condition really is and what our future prospects actually are, we must first understand the causes of our predicament. Only when we have done this can we hope to find ways and means of removing those causes and, eventually, of steering human life on a governable course.

It is quite evident that the interweaving factors that make the contemporary world unstable, unpredictable, and perilous are numerous. Basic among them, ironically enough, is the very position of supremacy over all forms of life that our species has attained on a planetary scale. As I will discuss later, our ascent to absolute dominance was perhaps too rapid for us to adjust to it fully; and the functioning of our worldwide empire requires political structures and regulatory systems of such unaccustomed dimensions and complexities that we have not yet

11

mastered the art of organizing and governing them satisfactorily. As a consequence of our incapacity to exercise our new power efficiently or lucidly, we are bewildered by all sorts of unexpected difficulties. We aggravate our situation by treating our new problems as if they were just another bunch of the old economic, social, and military difficulties, simple enough to deal with in a conventional manner, individually, each one on its own terms. This is patently not the case. Though we do not yet seem prepared to recognize it, in reality, we are faced by a new range of phenomena – and our problems stem from the malfunctioning of the human system as a whole. New mazes of problems ramify within one another and continuously interplay to form what The Club of Rome has termed the *world problematique.*

To seek distinct, discrete solutions for individual parts of this *problematique,* as much of society continues to try to do, is quite evidently a waste of effort. Everything is related to everything else: causes, problems and solutions are all interlinked in one great continuum. If we want to guide ourselves sensibly into the future, we must consider the entire dynamic picture of the globality of things, not merely examine some of its aspects in isolation. This is all the more necessary nowadays, when the very human condition – unique and lofty yet in dire jeopardy – is at stake. Neglect can no longer be condoned; mistakes cannot be tolerated.

The questions involved are, however, so numerous and so involute that we do not yet possess the maturity and the means to broach all of them from all directions. We must simplify our task and find an easier, though still appropriate, approach. In this respect, it is opportune to recall the warning of the great thinker and humanist, Teilhard de Chardin, who said that, whatever the importance of the matter being viewed, the angle from which it is viewed is no less important. The point or points of entry chosen to examine the current *problematique* as

12

thoroughly and comprehensively as possible are therefore of paramount importance. They cannot be dictated by purely economic considerations, even though these are the concerns to which we are inclined to attribute precedence. The economic aspects cannot be held to represent the entire *problematique,* since they either reflect mainly the interests of some particular nation, class, or culture or respond to exigencies emerging from contingent events.

The points of entry adopted to analyze the complex matter related to the state and possible fate of humankind at this turning point in history must be selected so as to permit us to penetrate to the very heart of the total human system and allow a truly global perspective. We can do this only by *giving preeminence to the relations between man and nature.* The profound changes these relations have recently undergone and the modifications that will certainly occur in the future are destined to affect human life more decisively than any other factor. Any discourse on the meaning of being human or on human prospects must first take into account how we stand in relation to the biophysical world, of which we form a part and on which our very existence ultimately depends.

Our perspective must be not only planetary in terms of space scale, but also very broad in terms of time scale. The dominant position our species has gained on the Earth is the result of a persevering process of multiplication, conquest, and coloniz-ation protagonized by our forefathers through countless generations and now being brought to a new climax by us. To envision the future stemming from the present, we must understand how our situation has come to differ from past situations and realize that the differences to come, which are likely to be even greater, will be differences more of kind than of degree.

It is useful to start back in the remotest ages and briefly consider the evolutionary developments of which we are the culmination – and in our opinion the all-important product. As far as we know, the Earth was probably formed as a celestial body upwards of six thousand million years ago, and life began on it halfway through this period. In order to put what has happened during these immense lengths of time in our small corner of the universe into perspective, we can compare the period with the days of a week, compressing the Earth's cosmogony and prehuman chronology into Monday through Saturday and leaving the whole of Sunday for the inception and unfolding of the last of the succeeding eras – *the extraordinary Era of Man*. If we assume that the Earth came into being in the first minute of Monday, life started to pulse early Thursday morning and from then on, little by little, unceasingly and tenaciously, wove its fabric, pushing ahead, spreading, evolving, mutating, and specializing into millions and millions of species and varieties. This process took ages. When they appeared, perhaps some 200 million years ago, the mammals were the most advanced specimens in the life-chain; Saturday evening was already fading away. Though life went on asserting itself relentlessly in ever more diversified and higher-grade froms, it still took quite a long while for some of the first anthropoids to abandon their forest sanctuaries for open territory and to roam around, attempting to walk erect and catch game for food. Eventually these anthropoids discovered that their hands had become free and could be used to perform newer tasks that stimulated the brain to organize itself better. With this development, the process of hominization had started. It took place some twenty, or perhaps ten, million years ago, at a time corresponding to between 11:30 and 11:45 Saturday night on the week scale. As this last sliver of time was passing, a big event was approaching.

On the stroke of midnight – about one million years or ten thousand centuries ago – *homo sapiens*, nature's last significant child, made his appearance here and there on Earth. Sunday had begun, and with it the Era of Man, whose first job was to struggle with the other primates and with all other creatures in order to start his own, amazing venture. It is still not yet quite clear whether this was a moment of glory or one of folly in the course of evolution whether, with man, nature created a lasting masterpiece or a freak the other forces of life would one day eliminate. At any rate, it is certain that, with man's entry on the scene, Sunday promised to substantially differ from any of the other days. Indeed, though, in terms of the week scale, at present we are at only two full minutes past midnight Sunday morning and man is thus still a newcomer among the other creatures, everything on our planet changed with his advent.

The new era launched by man is quite uneven and strange. It can be divided in two periods: *prehistory,* which lasted ninety-nine per cent of man's ten thousand centuries, and then *historical time.* During prehistory, although acquiring strength, man remained primitive and made things evolve at a relatively slow pace. Each new century must have seemed almost exactly like the ones before. Then, suddenly, about one hundred centuries ago, our species began making events move faster. Practically everything our ancestors did or recorded and bequeathed to us has taken place in these last one hundred centuries, namely, during the ten thousand years that we call historical time, which represents a mere one per cent of man's total era and corresponds to just one second in our cosmological reference week.

The five hundred generations succeeding one another during the relatively short period of recorded history have accomplished amazingly more than did their predecessors in the whole of prehistory. All this accomplishment began when man started cultivating a few patches of land and then went out to explore and populate the planet, strengthening his rule by founding colonies, trading posts, and finally mighty empires. The human spirit was elevated by the great religions man

15

conceived, and life was enhanced by arts and culture that flourished in dozens of civilizations. The inquisitive human mind was trained to learn more about the substance of matter and the mystery of life. Progressive discoveries eventually brought our forebears into contact with the infinitely small and the infinitely large in our universe. Human ascent, though, was invariably permeated by a deep sense of awe and reverence for nature, to which men paid homage either for the inspiration and bounties it offered or for allowing them and their families to exist. And, when they found new lands, plants, and rivers, devised more efficient ways to use the old ones to serve their wants and whims better, or learned how to navigate the seven seas more safely, human beings felt grateful and profoundly respectful for the overwhelming might and majesty of their terrestrial environment.

During most of these one hundred historical centuries, man's rise above his primitive condition took place at a pace that was very leisurely compared with what happens nowadays. Then things changed. The tempo of human events speeded up, and man felt stronger, less and less dependent on natural events and constraints. The newest phase of acceleration started about two centuries ago – a fleeting instant measured by the great clock of time – and has gained momentum on the wings of the industrial, scientific, and technological revolutions ever since. These *material revolutions* have given our generations fantastic and unexpected knowledge and power that can be – and indeed have been and are being – used for both better *and* worse. The human condition and human emotions changed entirely.

We now strive to take every possible advantage of the dominant position we have acquired, (which has fostered beneficial developments beyond anything our ancestors could have dreamed of) but we still are not satisfied. There seems to be no limit to the capricious, jumbled, irresponsible manner in which we use our formidable new assets to mutate the order of

16

things on Earth with the sole aim of asserting ourselves and promptly satisfying our greed. Anything that titillates our ego is apt to blind us to all other considerations and render us insensitive to the consequences of our actions. We fight for supremacy among ourselves and engage in cut-throat competition to reap quick profits, no matter what the cost to others or the possible infringement of ethical standards. While doing this, we devastate our environment. Indeed, the greatest abuse and the worst effects of our novel knowledge and power arise because they have made us so presumptuous and self-centered that we have forsaken communion with nature.

In a very short time, we have decimated innumerable species of animals and plants that, from the earliest times, were companions and provided us with support. We have degraded our environment by fouling and poisoning the very soil, air, and water on which, like all other creatures, we depend. We have built a mammoth technosphere of sprawling artificial cities, industries, and all kinds of artificial systems, heedless of the fact that the ever-increasing space and resources they need are acquired at the expense of natural systems and that their proliferation subjects us to the strains of an unnatural, mechanistic, and congested way of life. Now we are finally beginning to realize that we will be called upon to pay a very high price for having succumbed to the temptation of trying to reshape the Earth as if we were to inhabit it alone, when the planet is beautiful and generous precisely because many other forms of life contribute to making it what it is. To crown our work, we have learned to fabricate the most artificial of our artifacts: The Bomb, which can destroy everything necessary to all life, human life too, in the twinkling of an eye.

Though by no means exact, the comparison I have just made between universal time and humankind's career would not be appreciably altered by an error of scale; nor would any modification, by a factor of two, or even ten in the relations

17

among its phases, make any great difference to the overall perspective of the place of man and his present civilization in the overarching unity of everything. The usefulness of this comparison, on the other hand, cannot be underrated. It simultaneously shows the entire human era in a proper time framework and highlights the immeasurable chains of events and factors that have generated us and have unceasingly provided us with conditions indispensable to existence and success. Moreover, it stresses our belonging to a much larger dynamic whole, indicating that *our oneness with nature is the primary element of our being*, and warning us forcefully that anything we do to weaken nature or our bonds with it ultimately and unavoidably weakens us too. Forgetting this link can only lead to fatal mistakes. In other words, although man has carved himself a privileged niche in the world he claims as his own, his position is becoming precarious and may be completely ephemeral unless he stops acting as the insatiable, obtuse tyrant of the planet.

Only by viewing the human venture in this overall context can we appreciate how suddenly and radically our status has changed, how, for the first time, we must now behave as the fully responsible protagonists of our venture. The awareness that, willy-nilly, we are the paramount agent for change on Earth and that what we do henceforth will be the main factor affecting coming events, and our own future too, is crucial. Our future will be long or short, marvelous or disastrous, rewarding or misererable, depending in particular on whether we conserve or degrade the fabric of life throughout the globe and, more generally, on how we modify it by using our immense knowledge and power. Equally important is the perception that, for the first time, ours will be a planetary future in the sense that its alternatives, whether of self-fulfillment or of doom, are likely to involve humankind as a whole, not just some nations or regions, independent of one another. These facts show the new contours man's uniqueness has assumed since humanity acquired the capacity of broadly interfering, for good or ill, with the globe's life cycles and systems and of shaping, purposefully or unwittingly, its own destiny, not only that of many other species.

Considerations of both the fundamental, albeit forgotten, paramountcy of nature in our lives and our new capacity to alter our relations with it are so essential that their presentation in a variety of forms can enrich our vision, just as variations on a musical theme can underline the beauty of music itself. For example, it is always worth recalling that *man started as a weakling who was instantly on the defensive* and that his numbers were extremely small. For age after age, he subsisted in sparsely distributed, loosely-knit associations of families, where he slowly developed a shrewd mind and the manual skills necessary to build shelters and produce tools, artifacts and weapons. The ability to coordinate brain and hand equipped him better in the struggle for life, enabling him to protect his territory from enemies and himself from the hazards of the wilderness and the vagaries of the climate, and eventually leading to the formation of the first primitive societies. While this prehistoric stage of the human enterprise accounts for almost the whole of our existence as a species, things changed only about one hundred centuries ago, when luckier or more gifted human clans and tribes learned to raise plants and herbs and to domesticate animals so as to produce food at regular times and to preserve this food till the next season. They then settled more or less permanently nearby and established the initial villages. From these relatively secure bases, venturesome expeditions sallied forth to seek still more food, land, water, or salt, progressively eliminating or subjugating the weaker human groups who stood in their way. The process whereby ancient history gradually unfolded is witnessed by remnants of early monuments, sagas, legends, and traditions which survive to this day. Then, as populations became bigger and their needs and means grew manifold, the tempo and scale of human expansion changed too, and higher and higher civilizations blossomed.

The latest phase of this historical cycle of settlements, exploration, conquests and domination, is both very short and is still under way. It started with the formation of the modern nation states. In the name of God or king and fatherland, such states have been moved to carry their languages, flags, and laws to the remotest parts of the Earth; in addition this urge has caused them to struggle with one another and fight virtually more wars than it is possible to recall. This phase has culminated in the worldwide process of decolonization and the emergence of the last two empires in human history, entities which we call superpowers. For all its importance and turbulence, this phase has only lasted a couple of centuries, or just about two per cent of recorded history. This has been a period of many changes in almost all fields except, unfortunately, those of political thought and institutions, which have have not evolved much beyond the levels attained during the previous centuries, to which our time remains thus politically an appendix. Culturally, too, this is an age that belongs largely to the past, even though the whole of humankind has been shaken by the material revolutions that have finally skyrocketed it to completely new spheres of knowledge, power, and opportunity. This is the reason why, paradoxically, man has never been as much in danger as he is *now, at the peak of his power.*

Our love affair with these revolutions must be considered against the background of this political and cultural mismatch. Mesmerized by our power, we do what we *can* do, not what we *ought to* do, and go all the way without taking into consideration any practical *dos* and *don'ts*, or even the moral and ethical restraints that we should consider inherent in our new condition. The consequences of our misjudgment or irresponsible behavior are quite evident. We have vanquished numerous illnesses without reducing our reproductive fertility, with the result that world population is multiplying phenomenally. Today, in a time of quarrelsome, so-called sovereign states that lose no occasion to arm themselves to the teeth, the way we have developed military technologies means that all humankind is actually playing with fire. Hurtling on full speed ahead and indulging in our propensity for material possessions and consumption, we have dramatically swelled the global demand for goods, foods, and services. We have created artificial needs, artfully expanding the range of what is considered indispensable by constantly renewing fashions and designing products with built-in technical obsolescence. The only method we have devised to meet the surging waves of our rampant militarism and consumerism is to draw increasingly on the natural environment and to exploit, indiscriminately, the most accessible mineral and fuel deposits and all the living resources we can lay our hands on. Such actions irreversibly impoverish our unique, irreplaceable world, whose bounty and generosity are not infinite. Even if all the other adverse situations we find ourselves in today were to be alleviated, in itself, *our high-handed treatment of nature can bring about our doom.*

The situation in the biosphere, the closely interwoven system of life hosted by the thin film of water, air, and soil surrounding the Earth, is most delicate and distressing. Experience and logic tell us that, when a factor grows exponentially in a system – as human pressure does in the biosphere – the system is either strong enough to curb the anomalous factor or is overwhelmed and mutated, if not destroyed. Only recently have we reluctantly started postulating that humankind, with its relentless demographic and economic growth, might be becoming such a critical and intolerable factor, likely to lead to one or the other of these two negative outcomes in the essentially limited, sensitive context of life on our planet. If we are on the way to becoming such a destructive element, we must either spontaneously alter

21

our behavior while there is still time, or inevitably suffer from the disaster we bring to the entire globe.

A crucial question in this respect is whether our interference with the biosphere and use of renewable resources – those that tend to renew themselves and that we facilely believe will continue to do so – does not exceed their built-in regenerative capacity. The question, in other words, is whether today's thousands of millions of humans are not so intoxicated by their techno-scientific and industrial proficiency and so enticed by the lures of material welfare that they will extend their exploitation of world ecosystems beyond the critical point. No one today denies the impact that the expansion of human civilization has had on the texture of life on Earth or claims that the magnitude of the destruction, degradation, and manipulation man has wrought on the ecosystems does not dwarf anything that has occurred in the past. Nonetheless, though we already have enough disturbing indicators that human beings may now be going too far in their onslaught on those systems, very few people dare to admit openly that these attacks are becoming too reckless and are passing beyond effective control. The stark reality, however, is that this is what actually happens day after day all over the world.

For one thing, *wilderness*, which was once man's pristine environment and remains the very heart of nature and the cradle of life, is rapidly disappearing even from those regions where it still exists. Wilderness is of immeasurable value in purely cultural terms; but its loss may have still more serious, even fatal, consequences for our very survival. The world's genetic pool, built, diversified, and perfected during tens and hundreds of millions of years in nature's laboratories, is being inexorably disrupted and decimated; and, often, the habitats essential for the very existence and evolution of many species are being destroyed forever. Innumerable microorganisms and other relatively simple living systems, whose functions are irreplaceable for the dynamic, self-corrective balance of the biosphere, are simply being wiped out. Only our present-day ignorance prevents us recognizing that the stupendous, teeming diversification of life is indispensable for our healthy existence. Every plant or animal, herb or insect, however minute or modest, is in itself a microcosm. To survive up to the present day, all of them have had to face and have succeeded in solving, sometimes very ingeniously, thousands of biochemical and biophysical problems that we often cannot even formulate. If we study them patiently enough, we find that all life forms are incredible repositories of precious information. Extinguishing other forms of life is thus worse than burning libraries for, with them, we destroy forever a source of knowledge which may not exist anywhere but in their natural wisdom and experience.

The fates of the remaining *higher animal and plant species* are also totally at our mercy. Those we find economically useless will be remorselessly eliminated or indirectly brought to extinction because the environments and resources they require for life are requisitioned to serve human needs and whims. Others may be allowed to survive, only to be hunted for money or pleasure. The species that are considered valuable will be raised in captivity or domesticated and systematically hybridized, specialized, castrated, inseminated, and groomed to produce food, fiber, skins and timber. However, nobody knows whether, by living constantly under man's protection instead of being exposed to the natural competition and selection that prompt genetic revolution and fitness among wild species, these economically worthy animals and plants will not eventually be condemned to some kind of degeneration, or whether their resistance to pests and diseases might not be sapped. Our design of choosing, from the splendorous panoply of life, only a few species and varieties to live with us in self-imposed, proud, semi-artificial isolation may thus prove to be illusory and may ultimately be foiled.

Moreover, under present circumstances, demographic growth

23

and economic expansion will require an even larger and *more sophisticated technosphere*, which will inevitably be set on a course of competition and collision with the biosphere. The consequence will be a permanent and increasingly grave degradation of our biological environment. Our ignorance of the extent to which this is going to affect our own existence and the neglect that at present accompanies such ignorance can no longer be tolerated. It is therefore imperative and urgent for us to acquire more knowledge of the cumulative effects of our actions and our lack of environmental policies. In the meantime we must adopt some stop-gap measures to protect and conserve nature, even on a tentative basis, and establish some kind of temporary moratorium on our most patently harmful activities.

As a prerequisite, our long-term conservation policies should be inspired by *a new ethic of life* based on the recognition that any damage we cause to the planet's life-support capacity will boomerang on us and, more generally, that our future condition and quality of life will depend, to a degree never before imagined, on our attitude toward the other creatures populating the Earth. Everyone must realize that the creeping ecocide for which we are responsible nowadays, as a consequence of our rapacious or improvident daily activities, can destroy us as thoroughly as the nuclear big bang and that, if we do not soon change our behavior, such a heretofore unthinkable epilogue could occur when we least expect it – for instance, even before our global population reaches six or seven thousand million.

The environmental campaign must be conducted with vigor because of our contemporaries' stubborn reluctance to recognize the close relationships between the population-economy complex and human ecology. Let me digress briefly to discuss this issue. Reluctance to face reality in this, as well as in other cases, is often no more than escapism, an attempt to justify our current weaknesses, inaction, and excesses by attributing them to certain traits innate in human nature. It is argued that early

24

human beings attained dominance on the planet thanks to ruthless behavior dictated by selfishness, greed, thirst for power, intolerance and aggressiveness, and that these traits were derived from our apelike ancestors. This line of thought goes on to affirm that the very characteristics that enabled our forefathers not only to get the best of animal competitors, but also to eliminate feebler or less intelligent human groups or specimens, are engraved in our genetic code and therefore remain in us now, at a time when humankind has reached its evolutionary apex. In their view, even if these erstwhile useful qualities have now become negative values, we cannot get rid of them. This thesis thus postulates that, since we continue to be subject to primeval impulses inherent in the beast still within us, we are compelled to accept a fate beyond our control.

In my opinion, however, this is a facile, scientifically and morally weak, assumption that could be accepted only if and when it were fully proved that the current human crisis is indeed rooted in practically immutable *biological* characteristics of our species and is not caused by temporary, flexible factors.

I particularly disagree with any view which maintains that we are in trouble because of some fundamental deficiency inherent in our being and reject the thesis that, having given absolute priority to the development of brain power, we have become too specialized a species, incapable of adjusting quickly to change in a world calling for the greatest adaptability. According to this last interpretation, we are doomed by nature's inescapable rules, ironically sapped in our vitality by a peculiar human virus – intelligence. In contrast to these theories, I am convinced that our current crisis is not embedded in our nature, and hence ineluctable. However serious, it is a finally amenable *cultural* crisis caused by the civilization that evolved from the Judaeo-Christian tradition and later spread everywhere.

This civilization, of which we are so proud, not only idolizes man and exalts his mastery over the world, if not the entire

25

universe, but also condones practically anything he does to assert his primacy and justifies any means he chooses to use to attain this end. While theoretically affirming that man should be guided by noble principles based on a whole body of spiritual, ethical, and moral virtues and values, our dominant civilization has such anthropocentric and self-permitting cultural underpinnings that man is practically given the freedom to shun these principles whenever he finds it convenient to do so, or whenever they interfere with the pursuit of his mundane interests or the flattery of his own ego. In our day, the propensity to give priority to tangible benefits and material considerations ends up by pushing our humanness into the background while, logically enough, bringing to the forefront the object of our major current cult: *development* – whatever meaning we may attribute to this now almost magical word.

The crisis we are grappling with is therefore a crisis of identity and of goals. It has grown within ourselves and is both the cause and the consequence of the profound inconsistencies that render our thoughts and actions nebulous and unsatisfactory. It underscores the mismatch between, on the one hand, the new realities we are busily creating and, on the other, our understanding of why they have come into being and our ability to live with them. Thus it is that we rush into the future, using the most advanced means at our disposal, while our thoughts, emotions, policies and institutions remain anchored to a past that no longer exists. The dichotomy between our technological sophistication and our philosophical and behavioral obsolescence provides growing impetus for the forces that have generated, and now tend to perpetuate our crisis.

I shall return to this question later. At this point, it is very important to realize that our predicament is essentially cultural, not biological, and that it stems from inner imbalances and time-lags in our understanding of the implications of both the radical and rapid changes we evoke in our world and the as yet unsatisfied need to adapt to them. Although extremely alarming, this vicious circle of malfunctions is not irreversible. It can be both corrected and reversed. Therefore, we must not despair – for our condition is not hopeless. As we shall see later on, each human being has within himself or herself large reserves of comprehension, imagination, and creativity as well as a wealth of unexploited, even neglected, moral resources. These reserves represent untapped potential that can and must be systematically developed to repair the damage we have done to ourselves and our environment and to restore the lost equilibria between us. This will permit us, finally, to turn the situation around. To realize this and to act in accordance with it is both the main imperative of our time and the central message of the dialogue contained in these pages.

The plainest and easiest way of reestablishing our cultural balance and restoring a modicum of sanity to our thinking is to enhance our awareness of the simple, basic truths about our relations with nature. If firm cultural and behavioral foundations are established in the pivotal field of the ethics of life, this in itself will represent such a tremendous advance over our current state of unenlightenment and lack of concern that results will be easier to achieve in other fields. Then, little by little, a chain reaction will lead to the evolution of more mature attitudes and a more responsible society. It is therefore worthwhile delving somewhat more deeply into this topic, even at the cost of repetition.

It is clear that by fantastically extending our abilities and by fanning our expectations, the material revolutions have radically changed the course of history and have made human life more artificial, more complex, and more exposed to a whole new range of unnatural problems and perils. Less than two centuries ago, industry was a matter of no more than a few mechanical looms, steam engines, and other simple devices relieving a limited number of workmen from toil. Now, after having attained colossal, truly amazing proportions, the world industrial establishment is both the symbol and the secular arm of our civilization. Yet, formidable as it is, the industrial revolution has apparently created more demands than it is itself capable of satisfying or the world resource base is capable of supporting.

Coming hard on its heels, the scientific revolution broadened human knowledge enormously – though unevenly. We possess a dizzying amount of knowledge about the secrets of life, down to the genetic code, and about the laws of matter by which the universe is kept in orderly motion and by which all its particles – even the most minute ones – are bound together. In other fields of primary importance to human existence, performance, or happiness, however, we know much too little. Apparently, the amount of our overall knowledge doubles every seven to ten years; but this fund of knowledge is lopsided and will continue to be so. The so-called exact sciences take the lion's share of all scientific investment, while the moral, social and human sciences trail far behind. Nor are these the only distortions. It is estimated that more than ninety per cent of all scientists who have ever lived are alive today; but, to our shame, almost half of them are engaged in so-called military defense projects. Moreoever, since the people in general are unprepared to absorb, digest, and make adequate use of all the information and data available, only a relatively small part of our marvelous scientific bounty trickles down to the level of the world masses. Obviously, immense reordering of priorities and effort is needed to make science truly pure, as it should be, and to make its utilization exclusively beneficial.

Similar considerations must be made with regard to the technological revolution resulting from the application of this knowledge to everyday life. There can be no doubt that, at present, its greatest thrust is directed toward increasing the position, power and prestige of the well-to-do. Its impact is, however, so pervasive that technology has succeeded in

28

bestowing unprecedented welfare and comfort on people everywhere – even if in grossly unequal measure – while raising false hopes about its miraculous capacities. Inflated expectations for a new age of plenty and for easy technological solutions to most human problems have all but blurred our sense of reality, have created the myth of growth, have sapped our moral fiber, and have changed our attitude toward work, thus becoming largely responsible for the current crisis. Still newer technological breakthroughs looming on the horizon offer ever brighter promises that may, however, ultimately, prove to be only mirages or even traps. We may be tempted to change our lot by digging for fabulous Eldorados deep in the crust of the earth and the bed of the ocean; by employing genetic engineering to improve the quality of plants, animals, even human beings; or by robotizing and informatizing all our activities and shifting the staging areas of our military duels to intercontinental space and the upper atmosphere. However, we cannot escape from or solve our very real problems of ignorance, intolerance, inequality, instability and insecurity in any of these ways; and the world *problematique* will, if anything, become still more baffling and intractable.

To return to our theme, these evolutions and the myths with which we have surrounded them, have obfuscated our judgment and prevented humanity from perceiving the extent to which they have enslaved us and separated us from nature. The world they are imposing on us is no longer the world of natural cycles and forces that was familiar to our ancestors over the centuries and millennia. That world represented the foundation of their intuition, wisdom, values, and culture, and it is no less necessary to us than it will be to the next generations. Nature is being impoverished and humiliated. Ours has become a hybrid world in which the original natural elements are swayed and transformed by the nonstop flow of aritificial inputs for which we are all responsible. The resulting situations are frequently so convulsive and unpredictable that we are at a loss to keep pace with them let alone to adjust to them. That is why the more we change reality, as anarchically as we are doing now, the more we lose contact with it and *find ourselves estranged both from nature and within nature.*

Further injudicious human intervention in a world already profoundly transformed, manipulated, and polluted may have definitely, irreversible, degrading, and destabilizing effects. Even the best thought-out and most carefully implemented plans and programs that, theoretically, could assure safety and comfort to large populations must be reconsidered. Human systems in general lack the self-adjusting, self-healing, homeostatic qualities that endow natural systems with flexibility and adaptability in the face of change and crisis. They are therefore in constant need of supervision and regulation. However, supervision and regulation are difficult to provide because various human systems, created by different peoples and nations at different times and designed to serve different, even divergent, goals, obey different logics. They overlap, interfere, and compete one with another; and, as they are bound to grow ever larger and more complex and to spread more widely over the planet, the danger of clashes and collapses grows bigger and bigger. All this means that the ensemble of our artificial systems must be reconceived and managed in such a manner as to blend harmoniously with the world's natural systems.

Obviously, such an overarching task cannot be left to individual nations but must be the joint responsibility of all nations and of appropriate international or interregional groupings. It is saddening to note, however, that, at present, the possibility of achieving such a goal looks very remote and that, in many respects, though primitive, the naked savage dwelling in the early savanna and the nomadic shepherd searching for pasture at the dawn of history were far better interpreters, friends, and partners of nature than are we, their sophisticated barbarian successors of the nuclear and electronic age.

Only in recent years has some awareness of and pre-occupation with our overexploitation of the Earth's resources emerged, though often for the wrong reasons. The world's establishment is a case in point. By and large, it overlooks what occurs in the biosphere and is worried by the possible depletion of the capital stock of such nonrenewable resources as fossil fuels and certain ores – or of their more accessible deposits – because this may hamper economic growth. Nevertheless, it is gratifying to see an increasing number of citizens realizing that the real points of concern are the cumulative negative effects of all our tampering with, and damaging of, the tightly woven fabric of life on the planet. In other words, it is encouraging to register the new consciousness of the fact that our fortunes can be much more seriously affected by even minor disruptions in the Earth's life cycles and in what we like to call the renewable resources, than by the profligate use of the world's endowment of raw materials.

What, then, is the conclusion we must draw from these considerations? I think that one can sum up the situation by saying that, although at this stage we should know we must cherish and care for nature and protect it with all our might, we are still being carried away by power and lust, improvidence, caprice, and voracity. We ravage, poison, and foul the natural world. We wound nature in its very heart, the wilderness, and plunder its treasure chests. We flout its basic laws of strength through selection and diversification. We weaken its ecosystems, one after another, by manipulation and specialization. We do these and many other things without knowing whether or how our inconsiderate and wanton attitudes will backfire, downgrading the quality of our life, sapping the integrity and fitness of our bodies and minds, thereby making us pay too high a price for the material benefits we are reaping. Only now is it beginning to dawn on some of us that, one day, the

31

consequences of our attitudes may be disastrous. This awareness is not enough. What is needed is the full realization that we are unmistakably on a collision course with nature and that, if we do not change course, we are destined to be the big losers.

The time has thus come to make a thorough reappraisal of our present outlook and stance, even if this fundamentally challenges our faith in the material revolutions and the concept we have built of progress, wealth, welfare, and civilization in this epoch. New guidelines for our thinking and action are indispensable if we are to march safely and serenely into the future. Essential among them is the consideration that no other problem can be properly approached, let alone solved, no economic or social development is possible, no plan can be realistic and no heritage we wish to bequeath to our children can be effective, nothing can, indeed, be lasting until and unless we succeed in *reestablishing peace and harmony with nature.* Together with that of human development, this is *the basic imperative of our age* and one of the foremost conclusions to be drawn from our/reflections on the ascent of modern man to a position of exalted power and unparalleled responsibility on our small and vulnerable planet. All other considerations can only be ancillary.

Limitations Unrelaxed

IKEDA: The Limits to Growth, a book compiled by scholars and others and issued by The Club of Rome in 1972, had an enormous worldwide impact. Its severe appraisal of man's future with respect to resources, energy, food, and population growth forced people everywhere to change their attitudes toward industrial and techno-scientific advance. Nevertheless, a number of optimists continue to insist that science and technology will solve all problems. Indeed, the majority of those involved in contemporary politics, economics and technology adhere firmly to this view.

While striving to reduce the numbers of the unemployed, increase their military arsenals, and stimulate industry in their own lands, politicians continue to hold out to their peoples the dream of a richer society. Economists continue to try to invigorate economic growth – probably because development and growth in business are directly linked with support of their own social positions. Technocrats follow a similar course. To put the case against them sternly, people from all three categories are consuming and destroying the natural resources, beautiful natural environments, and essential conditions for a fuller way of life for all – conditions that ought to be the heritage of posterity – because doing so is an easy way to maintain and elevate their own positions. Paying homage to facile optimism and encouraging others to do the same are unforgivable actions because they are reprehensible in relation to future generations.

33

Sympathizers with the stands of overly optimistic politicians, economists, and technicians condemn indications of the gravity of the current crises on the grounds that they weaken people's will to grow and develop. In Japan, this attitude has led the Ministry of Education to request publishers of primary- and middle-school textbooks to delete pictures of the victims of atomic bombings as intolerably horrible, and to correct and adjust articles about industries that pollute the environment. The ministry is guilty of putting the cart before the horse. What they should be insisting on is the prevention of production, stockpiling, and use of the nuclear weapons responsible for the horrors they deplore in textbook illustrations. People who assume an optimistic stance in connection with polluting industries and the reckless consumption of the world's natural resources are guilty of similar folly.

PECCEI: In a nutshell, the first report to The Club of Rome affirmed that the rate of economic expansion prevailing in the world at the end of the 1960s and the beginning of the 1970s could not continue, simply because that growth was exponential while the planet on which we live is finite. And it warned that, if humankind nevertheless tried to force its way toward bigger production and consumption, powers beyond its control would some day stop it and probably bring its painstakingly created system to the point of collapse.

As you rightly say, the economists and politicians, as well as a good many industrialists and labor-union leaders who refute this reasoning and claim that the economy can be restored and maintained at the levels of its boom period, are moved by short-sighted self-interest. These advocates of growth at any price brush aside with one stroke the long-term interests of humankind and its present moral obligation to pass on to the future generations a planet that is in not much worse condition than it was when we inherited it from previous generations. They also refuse to admit that, in the middle of the last decade, a period of exceptional economic expansion came to an end and that the conditions that made it possible cannot be made to happen again.

For the time being, in spite of the efforts being made practically everywhere, there is no sign that the world economy can repeat its earlier extraordinary performance. The desire to see a strong upturn of world production and consumption matching the yearly growth rates of the past decades is thus likely to remain no more than a pipe dream. More realistic considerations must guide our thinking and policies. We must convince ourselves that, in the long run, some kind of sustainable balance between our overall demographic-economic dimension and the Earth's carrying capacity is indispensable and that it is up to us to find the ways and means to contrive it.

To return to the first report to The Club of Rome, there are sufficient grounds to consider that its central message holds good for the foreseeable future. Its discussion of the limits to growth has helped refine our understanding of these matters. The myth that growth – growth of any kind whatsoever – is good in itself had polluted our minds for a long time and is a misconception that still retains some power to tantalize. The Club of Rome report, rejecting the idea that growth for its own sake is a worthy goal, opened the way for the emergence of such new concepts as organic growth, sustainable growth and development that have helped steer our thinking in saner directions.

Energy Projects and Perils

IKEDA: Oil and coal – unfortunately major pollutants as well as important energy sources – are nonrenewable and will someday be exhausted. Many people expect nuclear energy to become a major energy source in the future. However, as the

35

numerous accidents that have occurred in various nations reveal, nuclear energy involves incalculable dangers.

The Three Mile Island incident in the United States stirred up worldwide reactions. Not long afterward, radiation leaks occurred in Japan – or it was publicly admitted that such leaks had occurred in the past. All of them were, it is true, outcomes of human error. But man errs by nature, and this means that the possibility of the same kind of accident recurring in the future is high. Drastically reducing the human element in such operations by means of extensive computerization will not solve the issue, since even the computer is neither omnipotent nor infalible. To date, no human life has been lost in nuclear-power incidents, but an occurrence of greater magnitude than has yet taken place could have horrendous consequences.

Disposal of nuclear wastes is another grave problem. No matter how thoroughly these products are sealed, there is always the danger that their containers might break or corrode. The scheme of loading them on rockets and propelling them to the sun is farfetched, at least for the present, since tremendous energy is required for such propulsion. Both of the two alternatives – burying in the ground or dumping in the sea – are accompanied by great risk. The peoples of Polynesia, as well as the Japanese, violently oppose the Japanese government's proposal to bury nuclear wastes in the Pacific floor. It is neither possible nor advisable to pin one's hopes on atomic power as an energy-source replacement for other fuels, since disposal of the waste generated by its production inevitably pollutes either the soil or the water of the globe. What is your forecast of energy sources for the years to come?

PECCEI: Energy, of course, is a major global problem, and how to approach it is still proving a controversial puzzle worldwide. Over the last few decades, humankind has been able to rely on hydrocarbons – particularly petroleum and natural gas – for its energy supply, actively exploiting and depleting the huge deposits formed during tens of millions of years in past geological eras. For various reasons, however, the situation is certainly going to change in the decades to come; we will no longer be able to count on these wonderfully convenient, but nonrenewable, resources to the same extent as we have in the past.

To start with, political difficulties may occur in our divided world. For example, whenever possible the oil-producing countries may have recourse to price increases or adopt conservation policies and restrict the output with a view to long-term phasing of the exploitation of this precious resource. Through the Organization of Petroleum Exporting Countries (OPEC), these countries successfully cornered the market during the oil minicrisis of 1973-74, trebling oil prices at a blow and putting Japan and part of Europe in a difficult situation. This was, to some extent, the aftermath of earlier improvidence. In the 1950s and 1960s, our industrial societies were inebriated by cheap, abundantly available petroleum, which could be easily piped or carried everywhere; as a result, they thought that they could rely on this bonanza almost forever. The awakening from this illusion was rude, but the shock proved salutary, for we now realize just how inconsistent we were in the sunny days of the economic boom, when almost everybody believed that the oil cornucopia was inexhaustible and ready to satisfy all human demands, however exaggerated these might be. Then oil and gas prices soared higher and higher; and, though in 1983 a buyer's market pushed them down somewhat, they are probably destined to increase even more in the not too distant future. The important fact is that people have become aware of the world's petroleum resources are not eternal; they may last another thirty or forty years, but not forever – and this compels us to research and search for adequate alternative energy sources.

Coal comes to mind as the first substitute because it is

abundant, but coal is far from being an ideal fuel, unlike oil and natural gas. Coal causes more pollution and is more difficult to extract and less readily transportable and usable. Moreover, the mining of coal is such unattractive work that is has become difficult to find people willing to go down into the mines. Projects are being developed to solve the problem by gasifying or liquefying coal *in situ* before pumping it to the surface, or even by burning it underground. While perhaps technologically and industrially feasible, however, these solutions are by no means waiting around the corner. The conversion of large parts of our economies from oil or gas to coal will be long, complex, and costly; and extended use of coal may cause such great – even if difficult to estimate – damage to the environment that it may have to be drastically limited.

Then there are the nuclear-energy solutions – a fad until a few years back. The case in their favor must be carefully examined. You hit the nail on the head when you pointed out the unresolved problem of safe disposal of radioactive wastes, which will grow to mountainous dimensions. Even if present-day electro-plants using fission technologies could be made clean, secure, and reliable, I am ready to argue that what is not reliable, secure, and clean enough is the human society that will host them. I am hard put to imagine how many of our societies, which are already in a great state of disorder and are incapable of governing themselves, could safely go nuclear or how, within the foreseeable future, they could find ways and means of living happily side by side with, and reliably protecting, several thousand huge nuclear stations dispersed in all their territories. And, even if this could be done, the problem of processing, transporting and storing quantities of deadly plutonium-239 (and other radioactive waste thousands of times greater than what it would take to kill all the world's inhabitants) would still remain.

At any rate, while the technical and economic problems of the nuclear society are immense and their solution not yet in sight,

38

still greater and more imperative are the political, social, ecological, and cultural issues they involve. One of these is the danger that the nuclearization of society may easily lead from the centralization of authority possibly required for the production of energy and policing of fissionable materials, to a concentration of political power that would be detrimental to the democratic life. In the face of all these objections, some people invoke the advent of an era of abundant, cheap, and safe energy produced not by nuclear fission but by nuclear fusion. This seems to me to be gambling prematurely on the possibility of making industrially operational something this is still at the theoretical stage. We should not bank on such pies in the sky. In any case, as things stand, it would be so hazardous for humankind to opt for nuclear solutions without having undergone a profound global socio-cultural and political transformation and having prepared its entire system for them, that such a course would be irresponsible to an unacceptable degree and therefore must be discarded.

IKEDA: The imperative step is to restrain consumption of energy resources. In spite of this need, however, people accelerate consumption by ever-increasing mechanization. For instance, the people of the industrialized nations do all they can to avoid actual physical effort in going from place to place. They rely on mechanical means instead of their own muscle power. This means they get insufficient exercise and must go to athletic clubs to employ more mechanical means to train. Both the reliance on mechanical transportation and the use of athletic clubs to get exercise that could be achieved simply by walking are wasteful. People could conserve energy by walking up and down steps instead of insisting on elevators and escalators. They would eliminate the need for athletic clubs if they walked short and moderate distances instead of relying on automobiles.

In combination with conservation steps of this kind, humanity ought to give serious thought and effort to taking advantage of the energy of the sun, wind, and water. Instead of relying on exhaustible sources like petroleum, coal, and the materials from which nuclear energy is generated, we ought to try to make efficient use of the operations of the natural elements. Current plans to do so are flawed because the sources of energy from things such as the sun or wind are unstable and the equipment needed to harness them is costly. Still, if man concentrated his knowledge and know-how on the problem, these difficulties could be resolved.

PECCEI: The availability of adequate, hence probably somewhat larger, quantities of energy than are consumed today is soon going to become a priority, calling for continental or regional – ideally global – plans for transition from a petroleum-based economy to one utilizing the greatest variety of energy sources. An essential tenet should be energy conservation in all possible forms, which means the development of policies, technologies, habits, and products that, while consistent with the requirements of a low-energy society, foster a high quality of life.

To put it in a nutshell, the fundamental imperative is to transform society so that it can do more with less energy and produce as much energy as possible from renewable, natural resources; namely, solar, geothermal, oceanic, wind, and biological. Originally, these alternative sources will be unable to satisfy our total needs; but our basic policy should be progressively to make them our main sources. Concerted research efforts for their development will no doubt show that, eventually, they may meet most of our energy requirements.

More Austerity

IKEDA: No matter what the energy source, I am convinced that

40

reduced consumption is a point of major concern. In our highly technological and mechanized age, people use far too much energy for the sake of comfort alone. Air-conditioning is an excellent illustration of what I mean. The wish to keep people efficiently busy, even in hot summertime, stimulates the use of air-conditioning. A desire for comfort leads people to use similar equipment in homes and even in automobiles, trains, and buses. This has a double ill effect. First, air-conditioning obviously requires enormous expenditures of precious petroleum as an energy source. Second, submission to constant air-conditioning lowers the resistance of the human body to illnesses like the summer cold. Our petroleum supplies are limited. By the time they have run out, will man be so debilitated that he can no longer survive in the natural environment? What is your view of excessive reliance on artificial controls on the world of nature?

PECCEI: Intellect sets us apart from animals; and, thanks to it, from earliest times our forefathers devised shelters, artifacts and weapons, which permitted them to live in relative comfort and security. Now, having grown much stronger technically, we have begun to exaggerate on all fronts – buildings, equipment, and armaments – thereby becoming accustomed to exist in an artificial habitat which demands little from the body, while placing increasing stress on the mind. All the marvelous modern conveniences such as air-conditioning, elevators, automobiles, servo-mechanisms, and so on may be pleasant and gratifying, but they have the effect of cushioning us from the natural environment. It is more than likely that, as you say, if we remain insulated from external challenges and outdoor hardships, our alertness will lose its edge; and our biological fiber and stamina will be debilitated. If for some reason we were forced to resume a more natural life tomorrow, we would feel utterly lost, because very few of us would be fit for it. I share your view that a reasonable dose of sobriety or austerity in the life style of all of

41

us who live in the so-called developed countries would not only enhance the quality of our current existence, but also better prepare us morally, psychologically, and physically for whatever vagaries or hazards life may have in store for us in the future.

Global Deforestation

IKEDA: Although disappearing rapidly in the industrialized nations, nature in the form of such things as forest covers persists in the so-called developing countries. Ironically, even there, the natural environment is being sacrificed in the name of development and economic independence. In Brazil, for instance, rapid deforestation of the Amazon Basin is actually causing concern about the global oxygen balance. The industrialized nations are in no position to order developing nations to abandon industrialization programs. Nevertheless, do you agree that they ought to try to do something about this serious problem through reforestation programs at home and aid abroad?

PECCEI: Deforestation is a very serious cause of concern, especially in overpopulated developing regions. Most worrying is the loss of tropical rain forests, both because they are the places where the greatest concentration of animal and plant species still thrives and because, once the tree cover is gone, topsoil is easily washed away by rain or hardened by the sun. Once a magnificent green mantle covered immense areas of our globe, renewing itself for countless millennia; but man disrupted its stable state, staking his claim for more and more of what it held in its fold – land and water, timber and firewood, food and skins, fibers and trophies. Forests began to disappear. Perhaps forty per cent of those extant when the first European explorers, conquerors and settlers began to infiltrate the whole world have already gone.

At present, primeval forest lands still exist in a few developing

42

countries; but, where strong conservation policies are not enforced, they are threatened with the same fate. In Amazonia, the world's largest verdant mass is being cleared away to make place for development, highways and settlements, or cut down for commercial exploitation. The situation in South East Asia and the Philippines, where large forest areas are lost every year, is similar. Elsewhere, a desperate struggle to survive pushes the poorest of the poor, who often rely almost entirely on firewood or charcoal for fuel, to destroy the trees they direly need and, when there are no more, to seek others always farther from home. Poachers, gold miners, and oil-prospectors hunting for treasures also enter and defile centuries-old forest sanctuaries. As this happens, the natural habitats of innumerable forms of life are destroyed forever, as I have already mentioned. And, instead of becoming appreciably richer, as they hoped, the countries and people of all these regions are actually becoming poorer, because their natural-resource endowment is being consumed, wasted and obliterated.

IKEDA: In developing countries, political stability and long-term independence are essential to counter these trends. Governments that are not sure of their duration in power concentrate on the profit of the moment and are unable to control and guide their populations away from unplanned felling of forests and such wasteful practices as burned-field agriculture. Stable governments, on the other hand, can engage in long-term economic policies and halt such devastation. For instance, I am particularly impressed with the way the government of the late President Park of South Korea was able to carry through an extensive reforestation program to restore a splendid forest cover to land that had been converted to naked red clay by unwise felling of timber. I suspect that similar success would be possible in South East Asia, Africa, and Central and South America. Political stability, ensuring a hopeful economic future for coming

43

generations would inspire the peoples of these regions to value their greenery. Furthermore, the abundant rainfalls and sunlight in these parts of the world would probably abbreviate the period needed for forest recovery.

The industrialized nations must provide aid to help these peoples attain political and economic stability. But, at present, such aid is not being provided in the right ways. The conflict between the United States and the Soviet Union has extended to these regions and stirred up war, with the result that the land is further devastated and made agriculturally non-productive. Ironically, the powerful nations whose antagonisms bring this situation about and who themselves look only to the profit of the minute, must then supply these nations with the foodstuffs they can no longer raise for themselves.

The whole world must become aware of the causes of this vicious circle and take prompt measures to eradicate them at once, for, if the destruction of the globe's forests continues at its present rate, the time limit at which there will be no wild greenery left on the Earth at all must lie not far in the future.

PECCEI: Impartial estimates warn that, at the present rate of destruction, the tropical forests will be practically wiped out in, at most, forty to fifty years' time. We are still unable to understand what this wanton stripping of the Earth's green cover may mean for human ecology and culture, but it would be foolish to believe that the consequences will be purely marginal. I keep asking myself why the cultured people of the world do not stand up as a body and cry in horror 'Stop the massacre!' Why do the churches, scientific communities, and the United Nations fail to warn governments and peoples forcefully that the retribution of today's forest destruction is tomorrow's mass human deprivation and death?

The situation in most developed countries, albeit not entirely satisfactory, is much better – thanks chiefly to the untiring efforts of the conservationist and ecology movements. Woodlands are protected; and, usually, where timber, pulp, and paper industries flourish, rational reforestation plans are obligatory. Let me note, however, that industrial reforestation does not solve the whole problem, because trees alone do not make a forest nor, when replanted, are they able to restore the often minute web of animal and plant life essential to the real forest environment. The reforestation plans of conservation-minded countries, however, are the best possible compromise at present, or at least a step in the right direction.

IKEDA: Unfortunately, in Japan, a highly mountainous nation, the forestry situation is far from ideal. The housing complexes which are spreading rapidly in the vicinities of major urban centers are resulting in extensive forest destruction. Tourism and development make demands for huge road-building projects that are devastating the habitats of vast numbers of different kinds of flora and fauna. Though ecologists, biologists, and animal lovers object vigorously, the people who own the land prefer to shut their eyes and ears and to give precedence to the economic advantages which derive from this kind of development. In other words, in some parts of the country the forests are protected, and in others they are ravaged. The environmental condition prevailing throughout the world is repeated in miniature in our nation, and probably in other nations as well.

It is unreasonable to ask poor people to renounce the right to improve their lot in the name of the protection of the natural environment. Yet, as we both agree, the natural environment must be protected. The only way out of this impasse is to devote maximum efforts to sponsor both environmental conservation and economic growth simultaneously. When this proves impossible, adequate compensation and assistance must be afforded to people, who must be relocated to less desirable places for the sake of protecting nature. Tax benefits might be

45

one way to do this on the national scale. Similar international assistance must be provided for developing nations when economic sacrifice in the name of the environment is unavoidable.

PECCEI: You have raised the question of whether it would be right to ask people living in an area that must be protected in the general interest to refrain from developing its full economic potential without adequate compensation. This question is pertinent both within nations and internationally. The general answer is that all reasonable efforts should be made to harmonize whatever development is possible with careful environmental preservation. When environment and development are mutually incompatible and the latter must be sacrificed in the interest of a wider community, reasonable compensation must be given to the people whose initiative is restricted. A mirror image presents itself at the nation level when, for reasons of national planning or overall land use, a potentially hazardous facility – such as a nuclear power plant – is to be located in a certain territory against the will of local inhabitants. Those people are entitled to receive adequate compensation.

The validity of this principle of equity cannot be limited to the national sphere; it should be made applicable internationally. If we are actually going to preserve global ecosystems from almost certain destruction, however, it is not sufficient merely to theorize on this matter. It behoves the strong, developed nations to take the political initiative in discussing, as a priority issue, for instance under the aegis of the United Nations, how principles of this kind can be implemented on a worldwide basis. It seems indispensable and urgent to me that any nation – particularly a developing nation – that might be asked by a community of other nations to adopt, in the name of regional or world interests, environmental measures that could restrict its own development, economic or otherwise, should be assured of fair

46

compensation. I know how complex and difficult it will be to reach agreement in the international arena on these principles and make them workable and enforceable. However I see no other acceptable alternative, short of a global ecological dictatorship, if we want to avoid man-provoked disaster and save our planet – forests, fauna, flora, wilderness, the atmosphere, and the climate, altogether – and our own skins too.

Italy is a case worth mentioning in connection with popular conservation movements. Incomparably rich in old monuments, art treasures, and scenic landscapes but blotted with one of the worst conservation records, she only recently rediscovered the venerable and profitable trick of making an asset out of beauty. A new movement aiming at intelligently preserving and exploiting her natural, historic and artistic heritage is gaining momentum. This movement is to be credited not to the official bureaucracies, but to a small grass-roots vanguard of citizens with vision, imagination and courage – environmentalists, naturalists, artists, poets, and a sprinkling of ordinary people, with school-children and the young in the forefront. It is thanks to them that a new sanity and ethic – *enlightened conservationism* – is gradually spreading among the people. Everybody, of course, wants economic development; but more and more people are refusing to accept it blindly, without counting up its social, ecological, and human costs. And, if any of these costs seem too high, the popular attitude increasingly insists that alternatives be openly discussed.

IKEDA: In Japan a number of conservationist groups are also active, but have not made splendid achievements comparable to those of their Italian counterparts, largely because they are hampered by deep-rooted bureaucracy. Though the Japanese bureaucracy makes important contributions to social stability and prosperity, it has the undesirable effect of squelching any movement that seems to violate established value criteria. A movement hoping to become strong enough to succeed in protecting our natural, historical and artistic heritage must arouse the awareness and obtain the consent of the entire people.

I find parts of Italy – especially the central and northern zones – so strongly reminiscent of Japan that I always have a comfortable feeling of having come home whenever I visit them. Yet, in terms of the preservation of the historical and cultural heritage, a number of factors make our two nations very different from each other. The splendid monuments left from the days of the Roman empire and the Renaissance in Italy are of stone and masonry. Traditional Japanese architecture is mostly of wood, which is damaged by water and fire. Consequently, the numbers of cultural monuments from our past are vastly smaller than the ones bequeathed to you Italians.

Furthermore, given the radical changes that have occurred in our life style in the past century, many Japanese no longer have any need of, or interest in, old works of architecture. Young people today prefer air-conditioning and central heating to the handsome, but undeniably inconvenient from the modern view, old houses of the past. Indeed, many of the surviving examples of old architecture have been declared important national cultural properties and have been converted into museums inhabited by no one. Conservation of the historic and cultural environments also demands understanding and a correct set of values. As you say, the green mantle of the globe and the ecological systems to which they are indispensable must be preserved if we want to save our skins. Similarly, our historic and civilizational heritage must be protected if we want to save our souls.

Halting the Hecatomb

IKEDA: Nature provides abundant offspring. An apple tree bears much more fruit than is required to ensure the continuation of the species. Birds are capable of producing more eggs than are actually essential. This abundance takes into consideration the risk of high offspring mortality. As long as he remained no more than just another normal predator intent on survival, man did no harm to the natural balance. When he let greed for wealth get the better of him, however, he began destroying animate and inanimate resources at a rate faster than nature can reproduce them. As a result, some species of fauna have already been driven into extinction and much of the world's mineral wealth is lost forever. Yet there is no sign of a letup in man's destructive pursuit of riches.

What is your opinion of my insistence that we must define and demand compliance with limits beyond which human beings must be forbidden to destroy animate or inanimate resources? Perhaps an international organization should be set up to study and act on such matters? Of course, the effectiveness of a program of this kind depends on the active support of all individual human beings.

PECCEI: I am very concerned with this problem, one aspect of which we have just discussed. Modern science and technology put at our disposal a wide information base and a great variety of means, from equipment to chemicals, that we employ mainly to exploit or simply decimate the Earth's living resources of flora and fauna; and, as you rightly note, this occurs faster than they can reproduce themselves. It is a matter of common knowledge, as well as being scientifically proven, that in many fields our destructiveness has outgrown the planet's biological fertility and that our pollution is outpacing the Earth's regenerative capacities. The stocks of the most accessible inanimate resources are being rapidly consumed. By these cumulative processes we are progressively undermining the very foundations of our own life, curtailing our chances of development

49

and well being, even of survival.

Particularly alarming is the hecatomb of other forms of life, ranging from the highest to the lowest, for which we are responsible. The guesstimate is that five to ten million animal and plant species exist in the world and that we will perhaps wipe out between half a million to one million of them before the end of the century. This is ecocide in the real sense of the word. Experience has shown that when, in an ecological system, one or more species disappear, the whole balance of the system is shaken and often degraded. Whether carried out under the pressure of urgent need or under the pretentious cloak of progress and development – no matter what immediate benefits some of us may derive from it – such continuous and widespread ecocide is something we and our children are going to pay for very dearly in the future.

We are making a mockery of language when we rhetorically refer to the oceans or the natural environment at large as *the common heritage of humankind*. In reality, we are all guilty of looting this heritage and of tolerating its desecration and devastation and the slaughter of all the non-human forms of life we and our fellow humans wantonly carry out. And we are all the more guilty because, as you say, each of us can, individually, do or make our communities or countries do something to stop it. Thousands of things of all kinds should be done and can be done if we really want to act.

IKEDA: Yes, many of the things that can be done are only indirectly related to environmental pollution and ecocide; and sometimes we fail to recognize them and their importance. People in the industrialized nations pay big prices for ivory, often without realizing that this inspires the indigent African to slaughter elephants indiscriminately since the tusks are an attractive and sometimes a unique source of income. If there were no buyers for ivory, there would be no need to kill

50

elephants. Political instability in the nations where elephant slaughter takes place deprives governments of the power to enforce anti-poaching laws. If the wealthy nations of the world would provide aid, leading to the stabilization of the political situations in these countries, effective measures could be taken to minimize wild-game killing. Such measures would include tracking down and punishing people who transgressed the poaching laws. However, in free economies, the rarer the article, the more desire is generated for it, and the more people are willing to pay to obtain it. This means that, no matter how strictly the law is enforced, there will always be people eager and able to circumvent it for the sake of high profits. Consequently, to stop such things as trafficking in the pelts, bones, and tusks of wild animals, governments everywhere must educate their peoples to the evil of purchasing, as well as selling, such things. Furthermore, local governments must provide their people with other avenues of profit and thus make trade such as wild-game poaching unnecessary, and limit the killing of animals such as elephants to the minimum needed to prevent overpopulation of the species.

I have concentrated on the issue of wildlife sacrificed thoughtlessly to provide poor people with a money income and to put luxury items on the markets of wealthy nations. This, of course, is only an illustration of the way apparently innocent acts – like the purchase of something made of ivory – can contribute indirectly to the devastation of world flora and fauna. I realize that there are many other acts that can be taken or avoided for the sake of saving our environment.

PECCEI: Yes, there are. I shall cite three examples of action that either is or can be undertaken – one aimed at reducing sea pollution and the other two relating to the respite that should be given wildlife.

The Greeks, who have been a great seafaring people since

51

ancient times and who currently possess about one-fifth of the world's merchant fleet, wanted to do something to reduce sea pollution. Their entire shipping community, from shipowners to seamen's unions, decided that it was high time for all their ships to stop fouling the seas while under way or in port. They asked me for The Club of Rome's moral support, as witness to their voluntary commitment to rigorous implementation of this anti-pollution campaign. As the Club of Rome has no organization, I called on four reliable environmental organizations to join it and the Greek promoters in studying a code of self-imposed conduct for safeguarding the seas, and ways of monitoring compliance with it. On Earth Day, 1982, this project was launched; and in less than one year it has received support from more than five hundred ships – an example that the rest of the maritime world is called upon to meditate and follow.

Returning once again to the case of my own country, the Italian public conscience is awaking to what must be done to safeguard the environment, but at present this is either not being done at all or is only being done halfheartedly. One instance is the growing movement to place a permanent ban on all hunting. In the relatively small and densely populated land of Italy, wild animals and birds have become so scarce and harrassed that, instead of being hunted and trapped, they must be given a sporting chance to reconstruct their herds and flocks if they are to survive at all. A change of heart and of laws is needed, and it appears that this is going to occur some day, mainly because of the pressure of public opinion. In your country, it is whaling that needs fundamental reconsideration. Could not our two countries, so advanced in many other ways, set an example and be each other's friendly, but stern, judges on hunting and whaling, with a view to agreeing finally on a mutual, honorable commitment to stop both, say, in three years' time? It is true that hunting in Italy and whaling in Japan are time-honoured activities deeply rooted in traditional culture, and that they

52

provide many people with jobs. But do you not agree with me that both our governments and our peoples ought to be guided by wider principles on such matters? Surely our nations are in a position to compensate all persons displaced or damaged by the cessation of hunting and whaling? Do you agree that a combined campaign should be started for the achievement of this twin goal?

IKEDA: As far as hunting is concerned, I am very interested to hear of the ban being considered in Italy and think similar regulative steps could be taken in Japan as well. As you may know, Japan officially protects certain animal and bird species that are known to be on the verge of extinction – the stork and the Japanese ibis, of which there are very few surviving specimens, are cases in point. But undeniably, each year, large numbers of wild birds and animals are slaughtered by hunters (who in their carelessness sometimes even accidentally kill or injure other human beings.)

Of course, strictly on the basis of the most scientific research, it is sometimes necessary to permit hunting of animals that are too numerous for their ecological setting or that – as the wild boar or the protected Japanse antelope called the *kamoshika* sometimes do – ruin agricultural crops, causing farmers hardship and loss. Nonetheless, even in instances of this kind, all possible steps should be taken to fence in or otherwise protect crops without resorting to shooting wild animals. And, when killing is inescapable, it must be – as I have said – limited and conducted on the basis of the most accurate scientific investigations. Hunters must be given moral as well as technical instructions in the handling of weapons, and no person who is psychologically or morally suspect should be permitted to own or use firearms.

I realize that Japan is constantly coming under international fire for whaling. No doubt there are many people in this nation

53

who rely on whaling for a living, and I do not think their needs and feelings can be ignored. Still, to prevent the extinction of these valuable sea animals, whaling ought to be stopped; and, as you say, the government should provide new occupations and financial assistance to people who are seriously discommoded by its cessation.

Procreation within Limitations

IKEDA: Population growth fundamentally influences conditions related to natural resources, energy and foodstuffs, which must be provided in at least minimal quantities to ensure the survival of the human race. By reducing infant mortality, improvements in and widespread diffusion of medical sciences have brought on explosive population increases, especially in developing nations. Though reduction in child deaths is welcome, in too many cases it is offset by sufferings caused by food shortages and malnutrition. Accustomed to having large numbers of children to ensure that some survive into adulthood, in these countries parents continue to breed to the limit, though modern conditions no longer make this essential. Rapid population growth is the result.

In order to guarantee that children born into the world have enough to eat and an opportunity for a good education, breeding must be limited. China and India have adopted government-sponsored guidance programs to this end, though, unfortunately, results have not yet equaled expectations.

Attempts on the part of authorities to interfere in this most private of all activities are unwelcome when they extend beyond encouragement of birth control and invade basic human rights. I consider such invasion even more frightful than material crises. What is your interpretation of the most suitable way to restrain population growth?

PECCEI: The ideal condition, I believe, is for each human being to be free to procreate according to his or her moral and practical judgment, while respecting his or her responsibility toward offspring and society. This means that, like all rights, the right to procreate not only presupposes obligations, but also has inherent limitations and should be interpreted and exercised by each individual with due regard to a higher, and more general and permanent, social, political, and economic good. Again, ideally, each human being should recognize these obligations and be able to adhere to these limits. Unfortunately, this is not the case. All too often love for children (when it is not a deep-rooted tradition), sense of pride, *machismo*, motherly instinct, fatality, or sheer ignorance override all these considerations, with the result that more children are born than family circumstances or societal health would allow for or advise.

In many parts of the world, attitudes concerning this problem are ruled by bigotry or demagogy resulting, among other things, from failure to realize the gravity, nature and causes of the situation. This provides still another instance of how bigotry and demagogy are the origin of endless human suffering – the helpless victims being in this case, more than anybody else, the children themselves. As well as individuals who procreate regardless of consequences, governments that believe ever-increasing populations to be consonant with their prestige and power policies are to blame for the explosion of world population and a consequent state of human affairs so dramatic that it may become tragic.

At the United Nations Conference on Population, held in Bucharest in 1974, a majority of states which advocated completely untrammeled freedom to procreate carried the day. Since then, several of them have at last changed their attitudes. One notable example is Mexico, which nonetheless still has rate of population growth so high that is has become the topmost national problem. The exponentiality of human proliferation defies our imagination and may push some countries to extreme policies. The Chinese case is typical. On and off, China has tried to control her rampant demographic growth, but with apparently unsatisfactory results. The 1982 census has finally and crudely brought to light the colossal dimensions of the problem, because population has more than doubled since Liberation Day in 1949 – the increase being from 500 to over 1,000 million in 33 years. At the same time, development studies have come to the conclusion that the country is capable of supporting a healthy population of some 800 million. Faced with these anguishing alternatives the Chinese government has chosen, if not been compelled, to adopt an inflexible policy of stern, even brutal, measures aimed at preventing the population from growing above the maximum target of 1,200 million by the end of the century – a very difficult objective to attain. This example demonstrates the terrible price that must be paid for failing to face population problems squarely and in good time.

IKEDA: The ability to feed is the crux of the limits that should be set on population growth. In the world today, some nations produce and export various manufactured articles for the money to import the foodstuffs required to feed their people. Other nations are largely the exporters of those foodstuffs; and, with the money they get from transactions with primarily industrial nations, they purchase the manufactured goods they need. However, the food-exporting nations, populations are growing and making increasing demands on home-grown foodstuffs. At the same time, manufacturing nations are in no position to demand food-exporting nations to limit their population growths so that they will have more food to sell.

When the exporters reach the point where they can no longer afford to export, the importers will have to learn to become self-sufficient. We must always remember that people live on the products of the fields and the seas; we cannot eat manufactured goods.

I realized that the matter of self-sufficiency is complicated. Different nations have different climates and therefore are only able to produce what thrives in their natural setting. This means that, even if nations do raise enough food to prevent hunger, for truly good health it will be essential to deal with other nations who raise the kinds of fruits, grains and other foods that are impossible to cultivate at home. Even in the Europe of the Middle Ages, a time when most local communities could feed themselves, spices were imported from distant Asia. Today, the populations of some nations have become so sophisticated in their eating habits that delicacies would have to be brought in to satisfy their palates. Nonetheless, after all this has been taken into consideration, it is still ideal to aim for basic food self-sufficiency and to limit populations to numbers that it is possible to feed.

PECCEI: No other issues are debated as controversially and as inconclusively as those linked with population. I do not believe that there is any such thing as an optimum population level, either in a nation or – even less likely – for the world as a whole. Too many elements come into play, including what standard of life the living generations – those who procreate – want to enjoy themselves and are willing to allow the next ones to have. What is necessary, however, is better knowledge of both the intricate relations between the population and its basic needs of food, housing, health, education, security, and so on, and how these needs and wants interact with the environment, resources, technology, institutions, and power structures, locally as well as throughout the global system. It is equally necessary to perceive that deeply ingrained procreation habits involve so many fundamental aspects of culture, values, beliefs and behavior, that to alter them will inevitably be a delicate, long and difficult task.

57

It is, moreover, worth repeating that all measures devised to this end must be inspired by the utmost respect for the human personality – respect for those who are now willing to give birth but no less for those who will be born – and that the complex web of human rights and obligations not only of individuals, but also of the whole human collectivity, present and future, must be considered.

These multiple requirements will probably present a permanent moral, political and social conumdrum. But they do not eliminate the need, in our present worldwide straits, for enlightened, effective family planning and population policies adopted by, or with the support of, the people themselves. And they should not obfuscate the implicit rule of life that no species can allow itself to multiply to the extent that its own existence might be endangered.

Food first, Industrialization Later

IKEDA: One of the many problems intertwined with the big issue of foodstuffs is the increasing numbers of nations that, owing to the international division of labor, are no longer self-sufficient in feeding themselves. In Japan, for instance, in spite of the paucity of arable land, argricultural ground is sacrificed to an unbalanced emphasis on industrialization. Though Russia was once a great food exporter and a fundamentally agricultural nation, today the Soviet Union is a food importer. Because of its deliberate attempts to stimulate balanced development between the agricultural and industrial sectors, China experiences no grave food problems at present; but, if the population continues to grow, the outlook for China's food future could be less than bright.

Today industrialization is considered the good way for productive organizations to follow. Undeniably, industrialization is essential from the standpoints of stimulating employment, increasing economic power, and strengthening the military. However, no matter how industrialized, all nations rely on their agricultural, fisheries, and animal-husbandry sectors for food. As I have already said, I consider food self-sufficiency essential if the nations of the world are to enjoy freedom from hunger. And, for this reason, the industries that produce foods – agriculture, fishery, animal husbandry, and so on – must get top priority. The industrialized nations use the money they gain by stimulating industry at the expense of agriculture, to purchase food from other nations that are still agriculturally oriented. Many of these bread-basket nations, following the lead of their so-called advanced neighbors, are now industrializing, with the result that the number of food-suppliers is constantly dwindling. Agriculture ought to be the most respected branch of any economy; ideally, all nations should strive for agricultural self-sufficiency.

In the face of the many starving people in this world, it is an urgent human duty to increase the available supply of foodstuffs. To do this, inefficient methods must be revised and improved. The sea may offer rich possibilities in this context.

Man may never be as successful in domesticating sea creatures as he has been with the creatures of the land. Still, pressing food shortages and scientific progress suggest that ocean ranches and farms may not be beyond the realm of possibility. Granted such projects become feasible, it is essential to ensure that man does not damage the marine environment the way he has damaged the land through farming and livestock breeding. What is your opinion of the prospects of the ocean as a potential source of foodstuffs? As you probably know, the Japanese have already achieved considerable success in the artificial cultivation of marine flora and fauna for food.

PECCEI: How right you are to put agriculture and food production, in all their forms, at the base of all human activities. This is in keeping with the tenet that nature and our relations with it must always come first in our thinking.

You mentioned, as an example, the oceans as a largely untapped source of food. This might well prove a promising field, though perhaps not to the extent that is expected. We received a warning in recent years when, in spite of increased financial outlays and more sophisticated equipment, the world's fish catch began to drop, an indication that the competing big fishing fleets are already overfishing the oceans. An improvement can hardly be expected without international cooperation aimed to put high-sea fishing on a sustainable basis. Such cooperation, however, does not exist, nor is it in the offing. More promising still is the idea of sea-farming in estuarine and coastal waters. This, however, is a craft in which the people of Japan, China, and the rest of the Far East in general excel, but which is less practiced or less ingrained in traditions elsewhere.

The problem of food for five, six, or more billion people must indeed have high priority in our thinking. To cope with it, new ideas and new approaches are essential. Basic, in my view, are the principles of regional self-reliance and global solidarity. The time when each nation needed to do no more than strive for itself, and when all were supposed to compete freely according to the mechanisms of a worldwide market has passed. Nowadays, food-deficit nations need to concentrate on self-help, while food-surplus nations have to beware of hunger abroad.

All major regions – particularly those of the Third World – should make concerted efforts to devise appropriate food strategies in which all countries of the region combine their efforts to become as collectively self-supporting as possible in this crucial field. To this end, production of locally consumed foodstuffs, and hence selective agricultural development and their indispensable underpinning of rural development, should be given top priority – if necessary, at the expense of industrial development. Food first, industrialization later. Although this guideline must of course be interpreted in relation to the situation in the different countries and regions involved, there can be no doubt that, if there is no food, there can be no industry either.

Even in this form, however, the goal of producing a reasonable food sufficiency for all cannot be attained everywhere. There are, for instance, areas in Africa and elsewhere that lack the required climatic and soil conditions, or that are chronically subject to drought or other calamities, that prevent them from growing enough food for their own needs. In such cases, it is the moral duty of the food-surplus regions – and particularly of the developed nations of North America and Europe – to come to their aid. It goes without saying that a broad-minded policy of this kind would be in their own enlightened self-interest too, for helping to solve the crucial food problems of the poorer regions would help to stabilize and improve the condition of the whole world system. In the long run, it will pay for the wealthier nations to consider our shrunken planet not as an arena where they are free to struggle to get the best of others and where, being stronger, they benefit from keeping weaker nations in subjugation, but as the common dwelling place we all must share, a place where all are interdependent and where each member fares well or ill according to the way all the others fare. Food problems may provide a good example to show that, by contributing something according to their own possibilities, each nation or region can increase the common good of humankind as a whole and can ultimately reap far greater benefits than by going it alone.

We know, however, that it may be easier to produce enough food for more and more people than to get it to the mouths of the hungry. Recent studies have shown that food shortages are likely to become dramatically worse in some regions, while some of the traditional bread-baskets of the world – like North

61

America – are probably going to become smaller than they are now. Nonetheless, the greatest problem will remain that of transport and distribution. This means that, even if reliance on foreign food supplies were justified and if money to purchase enough grain and the means to carry it to distant ports of needful countries were available, the most difficult problem would still remain trucking the incoming grain inland and organizing an adequate distribution system to get it to people who may be dying of want.

Broad cross-border solidarity and cooperation both within and among the large regions of East, West, and South are therefore indispensable not only for the production, but also for the mobilization and distribution of the food required by a multibillion humanity. Appropriate regional and interregional agreements, agencies, mechanisms, and infrastructure facilities at all levels will have to be established, and operated one way or another by the world community, if this primary need is to be satisfied. Its satisfaction is a precondition for peace and stability in our age.

IKEDA: I agree entirely. Insistence on at least efforts toward food self-sufficiency is a basic prerequisite for survival. To it I add the realization that some nations are, as you point out, incapable of raising the food they need because of the poverty of the soil. Furthermore, land that has great environmental, scenic, or cultural value in its pristine state must not be converted into farmland.

I object strongly to a trend, especially noticeable in Japan today, of converting what was once farmland into residential or industrial zones. Housing and factories can be built on land that is poor and otherwise unsuited to agriculture. Rich arable areas must be put to maximum use in food production. Rational use of land in ways that enable nations to maximize food production must be sponsored within nations and on a global scale. If what you call cross-border cooperation does result in the establishment of agencies and mechanisms for dealing with food shortages, such organizations must take this kind of land apportionment into consideration.

It is true that the warmer regions of the world house the vastest industrial establishments, but it is possible to envision a future in which largely automated industry is concentrated in deserts, and perhaps Antarctica. Such a situation would free the majority of mankind to live in the temperate zones, where they can farm, raise domestic animals, engage in commerce within a framework of respect for our great urban heritage, and devote time to the arts and crafts.

Desiccation and Deforestation

IKEDA: Natural elements not subject to human control – temperatures, rainfall, soil quality, and so on – determine the success or failure of agricultural efforts, though sometimes man's actions gravely alter these elements. For example, reckless development projects trigger serious alterations of the agricultural environment. The construction of the Aswan High Dam has dealt Egyptian agriculture a heavy blow. It is said that the grazing of domesticated cattle destroyed the grass and forest cover of ancient Mesopotamia, producing the aridity that reigns there now. Human beings hunted and killed the puma in North America to protect their herds of cattle, with the result that sheep and cows devastated the verdant cover of the prairies to contribute to desertification.

Any violence done to the delicate balance between flora and fauna invites the threat of biological destruction in what you have called the domino effect. In spite of this, however, man goes on destroying the foundation on which his own continued existence rests. When rains falls on forested land or grass plains, it is stored in the soil for a long time. Once the green cover is gone, however, rainwater either runs off or is quickly vaporized, leaving agriculturally useless aridity. In addition to population explosions, reduced rainfall and poor crops are given as reasons for food shortages in parts of Africa and Asia. For instance, desiccation and hunger have been tragically apparent in Ethiopia for a number of years.

The Sahara is said to have been spreading for several thousand years. What is the reason for this? Can anything be done? Does the situation irrevocably affect the food situation in Africa?

PECCEI: Apparently there are sufficient indications to suggest that in large parts of Africa, and elsewhere, desertification will get worse in the years to come. Some of the ensuing changes will be irreversible. You are right to mention the advancing Sahara. In certain areas, it has been moving at a pace of fifty kilometers a year. To some extent, the cultural attitudes of nomadic tribes in those places are to blame. Those people consider it prestigious to own as many cattle as possible; but, when kept in excessively large numbers, animals destroy vegetation from its roots. Badly planned agricultural settlements have aggravated the situation. Now numerous wells are being dug to water animals and raise crops, with the result that the water table is dropping and surface vegetation is dying, turning what had been a life-sustaining land into desert. This self-reinforcing vicious circle is difficult to combat because of the combination of traditional beliefs, inbred habits, and the sheer need of populations, plus the current universal urge to develop at all costs, with little thought for the consequences – all of which results in the gross mismanagement of land and water resources.

In other parts of the world, not shortage but over-abundance of water owing to man's meddling brings disaster. People living

64

in the Himalayas continue to fell timber for firewood and lumber. Depletion of mountain forest coverage in turn causes the uncontrollable runoff of water that is the reason for the ever-more-frequent, ever-more-damaging floods devastating vast areas downstream in India and Bangladesh.

It is not only population pressure and soaring demand on all sides for more of everything, but also the indiscriminate use of modern technical power in pursuit of short-term benefits that accelerate the pace of serious degeneration of our global environment. Besides those that you have mentioned, there are others such as the salinification and laterization of soils and their erosion by industrial civilization; the eutrophication of lakes; acid rains; accumulation of carbon dioxide, fumes, and dust particles in the atmosphere; the reduction of the ozone layer in its higher strata; and, in general, the environmental pollution, fouling, and devastation we bring about to reap quick dividends from all the natural resources we can lay our hands on. What appears to be a gain in our standard of living is really of no more than ephemeral value, and represents a permanent loss that we inflict on ourselves.

This thoughtless and senseless behavior is due to the abuse and misuse we make of our power; and the problem is primarily cultural. Experience has vividly shown that, if we leave the world to those who actually wield power – politicians, developers, technocrats, bureaucrats, and businessmen – we walk straight into trouble, leading with our chin, because these are the very people who give insufficient consideration to the indirect or delayed effects of their acts on our sensitive ecological and social environments. Engineers too, who in their own field can do splendid things, must be restrained. Today, as human presence and pressure are ubiquitous and the state of the world becomes increasingly degraded, we need as many poets, thinkers and educators as we need people of practical action. And, among people of practical action, perhaps we need fewer mechanics, truckers, and fitters and more farmers, agronomists, and irrigation experts.

New View of Our Place in the World

IKEDA: In East and West alike, primitive man saw other animals as possessing a spiritual nature and emotions very much like his own. Later, belief in human supremacy led man to despise other animals as instinct-driven creatures that he can exploit or even exterminate without compunction, according to his own whims and needs. The degree to which this occurs varies among species, but I see signs of emotional attachment and more than instinctive understanding in many animals, especially in the affectionate ties between parents and offspring or mated partners. Often, this affection is apparently more developed than in man. Animals have much in common with us and much to teach us. This is one of the major reasons for protecting them. What is your opinion of the emotional, and possibly spiritual, aspects of life in animals other than man?

PECCEI: I think that this question should be answered in two parts. Animals certainly demonstrate affinities, affections, and animosities similar to the ones we experience; and some are bound together by strong family or group ties, while most of them are instinctively led to sacrifice almost everything – even more than we humans – for the sake of their offspring. The fact that they share something important with us, or even outstrip us, should increase our respect for them. This is not the case at present.

The dominant contemporary belief, as I have repeatedly noted, is that the world and all life in it are at man's disposal. We are the masters. Nor is it exclusive to the Christian religious tradition to affirm that man was created in God's image – a

66

claim not far from man's own self-deification. That the world has been made for us, that it is unquestionably ours in all its aspects, and that we can do whatever we want in and with it have become such articles of faith everywhere that they are now embedded in the very foundations of our thinking. From these assumptions, it is a short step both to rejection of the idea that we, the lion, and the spider are parts of the same whole and to the assertion that, since we are superior to and different from other creatures, we rightfully have the power of life and death over them.

Even if these theses were acceptable, they could not be stretched far enough to condone our pitiless and cruel behavior toward many defenseless animals. We often act like savages, greedy despots and oppressors, not like sensitive and compassionate – humane – masters or trustees who understand and protect other creatures, even though they must use them to sustain their own life.

IKEDA: On many occasions, I have also said that the Judaeo-Christian concept of man as created in God's image is one of the causes of humanity's cruel and violent behavior toward other creatures.

In contrast to this stand, while valuing man as the creature capable of attaining enlightenment, Budhism recognizes the Buddha natures of all non-human creatures and, more important still, emphasizes the importance of compassion for all things. The story of how, in a former existence, Shakyamuni Buddha gave his own flesh to feed a starving tiger illustrates the extent to which Buddhism venerates compassion.

The Buddhist view is that man is created and sustained, not by a unique deity, but by all the casual and conditioning elements and things around him. Since he enjoys the blessing of interdependence with and support from all things, man must do all he can to return these blessings by caring for other creatures as kindly as possible. This, of course, is the idea behind the Buddhist injunction against taking life. Superior in many respects to other creatures, as you have said, man ought to be their caretaker and not their ruthless oppressor.

PECCEI: While there is no doubt that our incomparably greater intelligence and knowledge distinguish us from all other animals, that same knowledge should tell us that, the farther out in front we are, the more we should feel the supreme responsibilities inherent in our humanity. Behaving as we do, we belie our pretended ethical inspiration and foolishly impoverish and lay to waste what we consider our domain. Sometimes I ask myself whether our prideful belief that we have a predestined right to world mastery is not one of the major causes of the baffling crises in which we have thrown ourselves.

The other part of the question – whether there might be an animal spirituality – I do not feel able to answer. Off the cuff, I would say that, if we remain as self-righteous and callous as we are now, we ought to doubt our own spirituality before wondering whether, among the many things that the animals share with us, there might be a certain spiritual light. What seems unquestionable is that the time has come to correct our whole view of the world, and our place in it, including the entire range of relations between us and animal and plant life.

First Things First

IKEDA: Always fascinated by the unknown, in connection with which he has shown limitless curiosity, through such things as the voyages of the Age of Exploration, man has learned a great deal about what was once hidden in the lands and seas of this planet. Though he has penetrated only slightly below the surface of these mysteries, there is now practically no pristine place on the globe. In the latter part of the twentieth century, cosmic space appears to be man's final frontier.

The fascination with outer space is by no means new. However, only in the second part of this century has man evolved the technical know-how to generate the vast energies needed to project vessels beyond Earth's gravity and to equip them for cosmic flight. These new abilities and skills have evolved largely, however, not out of pure scientific inquiry, but in relation to the general growth of military technology. Spurred by a desire for national prestige the Soviet Union and the United States have vied enthusiastically with each other in this field. Their fervor seems to have cooled somewhat now, but could be fanned to flame again if a crisis threatened. Today space travel is prohibitively expensive, but someday costs may drop. What kind of gains do you think mankind can expect from efforts in this field?

PECCEI: In spite of the magnificent achievements of space exploration and the scientific progress they have fostered in related fields, I do not think that the present generations and the next ones should expect dividends of the same order of magnitude from them. Further knowledge and technological advances will definitely result from new space programs; but the programs themselves will demand enormous amounts of talent, means, and money that could produce more beneficial results if applied in some other way.

Yet, I do not think that man will, or should, ever give up the idea of going into space to search for something new or to conduct research on something about which he wants to know – even if his motives are not all highly principled. Three major motivations are said to push him onward at present: the fascination of mysteries beyond, the lure of possible unknown resources, and the hope of newer military footholds. Undeniably, the military objective is the strongest, as the United

States and the Soviet Union have amply demonstrated by spending billions of dollars and rubles to get a strategic edge on one another in space. But perhaps there is a fourth reason too: the illusion of being able to escape from the problems of this Earth. Why do I believe that, even setting aside all military designs, our space ventures will continue? The simplest answer is because space is there and we want both to know all about it and to demonstrate our presence wherever we can.

These attitudes and our space enterprises, however, are acceptable provided that they do not exacerbate our rivalries or belligerent impulses and that we do not forget that our home base down here on Mother Earth is full of unresolved problems requiring our greatest attention. Paramount among these problems are the ones engendered by our dreadful relations with Earth's nature, our parent and indispensable, permanent source of life. To redress these relations must be our foremost objective. First things first.

PART II MAN AND MAN
Views of Daisaku Ikeda

II – MAN AND MAN

Views of Daisaku Ikeda

Man is his own most intimate environment. From birth to death, human beings in today's advanced societies spend practically no time out of reach of other human beings. In infancy we live in contact with mother, father, brothers and sisters, who teach us how to talk and help us acquire various kinds of personality-forming knowledge. As the individual grows a little older, he spends most of his time with teachers and schoolmates. Once a member of adult society, he must relate with co-workers and clients directly and indirectly connected with the services or products his work involves. The creative artist – ostensibly the most independent and perhaps isolated member of the human race – must relate to others. First, he must learn skills from teachers. Later – and this is particularly important today – he cannot afford to overlook good relations with clients who buy and appreciate his work.

More or less direct contacts of these kinds do not, however, explain the entire human situation, for each individual human being depends on vast social mechanisms – administrative, production, and communications networks – complex interlockings of human relations that sustain his very existence. Though in one sense they restrict him, in another they convey great benefit to man.

For these mechanisms to function effectively, subordination is essential; that is, some people must be able to set goals, plan, and direct, and others must put into action directions handed

73

down from above. In this setup, an inevitable outcome of the activities of the social organization, rulers monopolize the right to make decisions, while the ruled lose freedom to participate in decision-making. This generates inequality. In addition to enjoying the other various advantages of this inequality, rulers can revel in the freedom of compelling others to act as they wish them to act. The ruled, on the other hand, must suffer the knowledge that they are deprived of the right to exert their will.

The interesting point in the ruled-ruler relation is the enjoyment the ruler experiences. Though, as I have said, this is a byproduct of the evolution of social structures, it has tremendous appeal for human beings. According to Buddhist thought, the desire to rule is the apex of desires and its satisfaction the paramount satisfaction. It is assigned to the final and paramount realm of desire gratification, which is, however, ruled over by the King of Evil. In short, Buddhism recognizes the evil in satisfying the desire to subjugate other beings to one's will.

Evil deprives others of pleasure. Life is man's greatest pleasure. Therefore, rulers who wage war, depriving their subjects of life, commit maximum evil. The social systems of certain nonhuman creatures – ants and bees, for instance – involve an element of subordination. Violators of the established social code are punished – though not executed as is the case in much of human society. However, I doubt whether the leader of any group of other creatures ever organizes his following with an eye to attacking, overcoming, and exterminating another group. Even the soldier ants who sometimes kill each other probably do not act on a directive from a ruler.

In contrast, human beings with no personal enmity mercilessly slaughter people their rulers arbitrarily designate as enemies. For example, during World War 11, many Japanese who had never harbored dislike for Americans before, suddenly fought them with loathing and animosity because the militarist government said the United States was an enemy nation. When Japan lost the war, overnight, while fearing them as invaders, the Japanese welcomed the Americans as friends. Though the sudden change in attitude can be attributed to the facility with which Japanese adapt to circumstances, in fact, it demonstrates something deeper and more generally applicable about the relation between ruler and ruled. It shows how fundamental submission to the ruler's will can be.

Though differences of degree exist, all man's attempts to form social structures inherently involve the evil of authority and power. This is a double tragedy. The tragedy of the ruled is that they must undergo suppression, often oppression. The tragedy of the ruler is dehumanization by power. The more sophisticated the society, the more suppressive tendencies grow. This means that, as long as human beings strive to elevate their civilizations, the magnitude of crises possible through evil in the human mind inevitably grows.

Love and Compassion

Of course, subordination is not the only relation which exists among men. The human mind is capable of love and compassion – probably the most beautiful attitudes in the animal kingdom. Christ emphasized love and elevated it to the level of religious psychology. European philosophers have pondered and argued the nature of love, which secularization has reduced to the ordinary relations between man and woman. I believe that, though he certainly used it to infer the bond between man and God, Christ meant the word *love* to stand for the most powerful tie possible between and among human beings.

Buddhists employ the concept of compassion to mean something like the Christian idea of love. The Japanese word

75

signifying this concept, *jihi*, is a rendition in Chinese characters of two ancient Indian terms: *maitri*, the giving of pleasure, and *karuna*, the elimination of suffering. When a father teaches his child skills and knowledge that will enable the boy to live a richer life than he, the father, has enjoyed, he is giving pleasure – even though it may look like very long-range pleasure. The mother who spends sleepless nights nursing an infant with a fever and suffering more than the patient is eliminating suffering.

According to religious doctrines, such things as love and compassion are attributes of the divine, of God and the Buddha. In my opinion, they are innate to human beings as well. Men have been known to abandon happiness or sacrifice their lives for the sake of others. Parents have willingly died to save their children. Others have leapt into raging seas to rescue perfect strangers. Young revolutionaries have died in the cause of liberty for their people.

Love and compassion may take apparently contradictory forms. On the surface, helping a weaker individual looks like compassion, when in fact it may have the adverse effect of making the recipient weaker and less self-reliant than before. On such occasions, refraining from the ostensibly compassionate act constitutes a profounder, more subtle compassion. It is compassionate to respect the rights of other individuals. It is folly to respect the right to liberty of a dangerous criminal who could work harm on others if allowed to roam loose. Matters of this kind obviously must be resolved on the basis of the mores and laws of the society in question; but actions taken toward their solutions must be based not on lust for power or self-interest, but on the kind of love and compassion that give Christianity and Buddhism their importance for man.

All forms of life concern themselves with self-preservation. Instinctively, both mentally and physically, living creatures flee the danger of death, which they can detect beforehand, in order to be quicker in putting themselves out of its way. In other

76

words, we are all fundamentally organized selfishly for our own preservation – and advantage. This organization is innate. Love and compassion are also innate.

In their immediate range of application – self, family, loved ones, and so on – innate love and compassion are limited. To enhance their value they must expand their range of applicability. They must never, however, lose their instinctive natures. Without sound, deep roots in innate instinct, love and compassion are empty if beautiful ideas, too weak to withstand the powerful onslaught of selfishness.

While remaining instinctive, they must be directed equitably to all people. To make this possible, we must come to consider all members of the human race as one family. Religions advocating universal love and compassion for all living things strive to make men aware of the general commonality of all life. The founder of Christianity, Jesus of Nazareth, called the Son of God, taught love for all mankind. The founder of Buddhism, Shakyamuni, who is called the Buddha (Enlightened One) because he attained an understanding of the fundamental Law of all life, taught that compassion should be afforded to all people with the impartiality of rainfall.

Patently, reinforcing the strength of such altruistic attitudes as love and compassion is the only way to halt human selfishness and lust for power. How can religions advocating love and compassion triumph over the annihilation unbridled greed threatens? What are the conditions essential to the effective operation of such religions?

Religion and World View

I should like to discuss the Buddhist viewpoint in connection with compassion for all humanity, not in an attempt to prove the superiority of any particular religion, but because, as a Buddhist,

77

I have a certain amount of knowledge on this topic. Buddhism recognizes a strict, ineluctable law of cause and effect in all life. This law applies to all the phenomena of the universe.

For example, a lawyer is a lawyer because of a variety of causes and effects. The effect is his present occupation and qualifications for it. The causes include, but are not limited to, talent, ability, academic career and success with the bar examination. I have said, 'include, but are not limited to' because not everything can be explained on the basis of causes evident in the present life. As I have explained elsewhere, Buddhism posits a universal life that manifests itself over and over again in countless rebirths. Each manifested, individual life has a karma account carried over from previous lives. Though they are certainly effects, such things as personal appearance, personality, physique, domestic environment, and so on are difficult to explain on the basis of causes limited to the present life. Buddhist philosophy explains such effects as the outcomes of causes from previous lives, not all of which have necessarily been lived on the planet Earth, since there may be innumerable other planets where life can manifest itself.

Since each life has been repeated over and over from the infinite past throughout the universe, it is entirely possible that at one time or another somewhere or another, all individuals have actually shared blood relations with all others – in short, in the strictest sense, we are all conceivably members of one universal, eternal family.

I have already explained that love and compassion are innate in life and related to the instincts – notably the instinct for preservation of self and species (family). The Buddhist ethic of compassion for all living beings is rooted in the instinctive desire to preserve this species, since the possibilities for family ties among all human beings in limitless numbers of former existences mean that all people are literally one.

Shakyamuni said that all the world was his own and all the

people in it were his children. Nichiren Daishonin, whom I and my fellow believers consider the Buddha and savior of the world, said, 'All the people in the world are my children. All of their sufferings are my sufferings.' A person who calls all mankind his children cannot fail to suffer when human beings suffer.

Family ties alone do not preclude hatred and animosity. Brothers and sisters can hate each other all the more virulently for being brothers and sisters. But, generally speaking, serious warring within a family is less common than belligerence between unrelated people. The important thing is to learn to wish for the happiness of others and to expand one's love and compassion to universal limits. The Buddhist idea should make this easier to do and should thus provide a sounder basis for a peaceful way of life.

The Judaeo-Christian myth of Adam and Eve – a doctrine of the brotherhood of man, since we are all said to have descended from this one pair of progenitors – claims that man was created, together with all things, at a specific time by the hand of God. Instead of positing a moment and agent of creation for mankind, Buddhism argues that human beings have existed since the infinite past (the likelihood of life on any of the other possible worlds in the universe answers the objection that such an idea runs counter to scientific common sense of biological evolution). I find the Buddhist analogy between the cause-and-effect nature of all life-phenomena and the mutual interrelations among all human beings more compatible than a mythological explanation of human origins.

Some recent developments in Western medical science and psychology give credence to the Buddhist idea of preceding lives. For example, under hypnosis a woman has projected her memory into the past to the extent that she recalls having been a housewife in France centuries ago. Historians have verified her language and the descriptions of her way of life in the past. A supernaturally gifted American has investigated past lives of several ill people and has discovered prenatal causes of their present illnesses. These case files provide fascinating research material. Nonetheless, it is difficult to provide objective, scientific verification of things that may happen after death or have happened before birth. Though hypnotic research into past and future lives may not find acceptance in the world of learning, it can make the Buddhist interpretation of life more convincing.

War and History

The most wicked of all relations among men is war. Though Buddhism in the East, and Christianity in the West, have attempted to teach the spirituality of humankind, in both parts of the world man's history has been a bloody pageant of warfare and misery. After having converted to Christianity, the Roman Empire collapsed under waves of tribal movements from the north. For centuries, Europe suffered depredations by incursions from the Islamic countries, from the Magyars, and from the Scandinavian Norsemen. Even after the pressure from without eased, squabblings from within among feudal lords plagued Europe. Nor did the founding of such national states as England, France and Spain put an end to the bloodletting. Quite the contrary, the establishment of monarchical states led to increases in military force and improvements in weaponry, both of which made war more cruel and brutal. In the early period of their history, the Europeans could at least claim that they were fighting in the name of Christianity when they battled invading infidels from outside areas. They had no such rational justification for the later battling and killing that took place among Christian nations, and which pitted members of the same religion against each other, people who ought to have respected each other as brothers in faith. Instead, however, they engaged in probably the most irrational and cruel warfare ever to have taken place. Awareness of the horror such combat caused finally led to a partial separation of religious faith from matters of state and war. Warfare fell exclusively into the hands of political powers, under whose auspices its scale and brutality have steadily increased.

In the light of its fundamental teachings, the Christian church ought to have been a torchbearer – at the time, the sole torchbearer – of hope for resistance against political authorities seeking to obtain ever-intoxicating power by converting ordinary human beings into cannon fodder. Ostensibly the inheritors of the justice of Christ, church and clergy alike became the pawns of political authority and even waged wars against their own. Now separated from political authority, the Christian church has lost the power to speak out against the tyranny and cruelty of war and the political might that brings them about. Even when it exercises what power it has in this direction, its words no longer carry much weight.

Has Buddhism been more successful in preserving peace in the Orient? In the third century BC, the famous Indian king Ashoka, who brought the majority of the land under his unifying sway after his own wars of subjugation, became aware of the cruelty of military control. Out of reverence for the spirit of Buddhism, he attempted to rule according to its teachings. Sacrificing himself for his people, he advocated respect for all life and built hospitals for animals as well as for human beings. He sent emissaries of peace to many foreign lands – it is recorded that his messengers went as far abroad as the capitals of Egypt and Greece – and concluded pacts of amity with many other nations.

In terms of religious inclination a parallel can be drawn between King Ashoka and Louis V11 – Saint Louis – of France. Nevertheless, the difference between their interpretations of international relations illustrates an important distinction between the Christian and Buddhist attitudes toward the value of life. Whereas Ashoka abandoned military force and sent out messengers of peace to his neighbors, Louis took up arms and marched on Jerusalem to take back Christian holy places from the Muslims.

Their successors also set the two kings apart. After Louis V11, France flourished to become a leading monarchy – though ruled by families other than Louis's own – until the French Revolution. King Ashoka's dynasty failed shortly after his death; and, unfortunately, no kings willing to rule in his spirit emerged thereafter. Though other Indian kings – for instance, Kanishka – caused Buddhism to flourish, no political authorities ever again ruled in a fashion permeated with Buddhist compassion.

In Japan, Prince Shotoku (574 – 622) and the emperor Shomu (reigned 724 – 49) were devout Buddhists who tried to reflect the spirit of their religion in their ways of governing. Though they are much less famous and operated on a smaller scale than he, they are comparable in terms of purity of aim with King Ashoka of India.

In 794, the capital of Japan was moved from Nara, where it had been when Shomu reigned, to the site of the present city of Kyoto. Thereafter, for three centuries the death penalty was abolished, and the nation knew neither internal nor external war as a succession of Buddhist emperors reigned. The rise of the warrior class, however, put an end to the long era of peace. A complex set of circumstances including natural disaster, economic instability, psychological unrest, and social change brought about this state of affairs. In terms of religious faith, the neglect of the *Lotus Sutra*, the Buddhist classic that most thoroughly advocated respect for life, certainly played a part in the debacle.

When the capital of the nation was moved to Kyoto, the emperor of the time had for his religious guide and teacher the

82

famous priest Saichō (also known as Dengyo Daishi), whose headquarters was based in a group of temples on Mount Hiei, not far from the city. Saicho's Buddhism is based on the *Lotus Sutra*. During subsequent years, however, contacts with China introduced other Buddhist sects whose novely captured the minds of many people, leading them from faith in the *Lotus Sutra*.

The *Lotus Sutra* is the sole Buddhist classic which teaches that the Buddha nature is inherent in all beings to an equal extent and that all human beings are therefore able to become Buddhas. In contrast to this doctrine, some of the newly introduced sects taught that Buddhahood is impossible in this world and that the only way to attain it is to die and be born again in a paradise (Pure Land) in the west and there to undergo further discipline. Since salvation in this world was denied, people disregarded what happened on earth. Such negligence opened the way for injustice, irrationality, and even cruelty.

In short, though Budhism has reflected its compassion on certain periods of history, in the overall view, oriental history has shown how difficult it is for societies to accept the true Buddhist spirit. Even when it has been officially established, under the intoxicating control of lust for power, people have forgotten it all too soon.

The National State and Peace

The dual role of the national state is to govern its citizens and conduct foreign diplomacy. When diplomacy fails to resolve problems, states resort to armed conflict, thus becoming the most virulent cause of mutual massacre in human society. All too often, domestic government looks like nothing more than preparation for military belligerence. Such was certainly the general policy of national states until after World War 11. A nation's crowning glory was its generals, its victory-bringers. Of course, vast quantities of blood and countless lost lives were the price the ordinary people paid for the glory of triumphant generals. National states – especially modern ones – could be called immense machines assembled for the sake of ever more efficient slaughter.

The Chinese were the first to evolve the sophisticated, efficient organization seen in modern states. Chinese ability to govern a sprawling land inhabited by many races over many centuries – in spite of occasional dynastic changes – bears witness to great genius in this branch of human activity. However China was never surrounded by efficiently organized rival states for on her borders were loosely grouped nomadic peoples or weak tributary states modeled on, but inferior to, the Chinese pattern. In other words, China rarely faced fierce rivalry or war with peers.

The situation in Europe has been entirely different. The several organized states that have constituted what is called Europe in modern times have engaged in ferocious warfare over and over again. World War 1, the most tragically costly of all human conflicts, showed the scale and awesomeness of the threat posed by warring national states. At the conclusion of that conflict, people were still insufficiently aware of this threat. Intoxicated with glory, the victors acted in a way that precipitated the second act of the drama of horror and folly: World War 11. This time, owing to grouping into entire camps of opponents and to the application of scientific and technological knowledge to weaponry, the scale of the massacre and destruction grew greater than ever before. Essentially cases of winners punishing losers, the Nuremberg and Tokyo war-crimes trials, conducted after World War 11 in the name of a humanitarian justice transcending legality, nonetheless provided chances to re-examine the formerly accepted concept that, no matter what happens, the state is always right.

Developments in the world since 1945 have weakened the hold of the idea of the national state on the mind of man. For example, improved communications technology has made possible the establishment of multinational business concerns operating throughout wide areas crossing several boundaries. Such concerns are welcomed by host nations for the economic stimulus and additional employment opportunities they provide. They also help to overcome the notion that people from other lands are uncomprehending, incomprehensible outsiders – and thus make humankind a little less parochial. The breakdown of colonial empires has created a more generally international outlook. After the exhausting experiences of two world conflicts, European colonial powers, who had oppressed many under-developed regions, lost their old holdings, where new feelings of self-determination inspired the formation of independent nations. Nevertheless, political, economic, cultural and other bonds have continued to link the new nations with their former colonial overlords and have helped prevent alienation and isolation.

Other factors have assumed greater importance and have helped break down the barriers of nationalist exclusiveness. Among them are the desire to secure natural resources, trade and commerce in raw materials, the efforts of peace-supporting organizations, information exchanges, and systems for cooperation in fields of scholarly research.

Unfortunately, however, fading national awareness and weakening of the power of controlling authority have not given love, compassion, and mutual respect top prominence in human relations. Lust for power, which has simply shifted from national states to other organizational bodies, continues to oppress mankind.

This is true because man in general has not yet awakened to his own inner nature. He cannot, therefore, act on the basis of the good parts of his own best inclinations. Lacking the knowledge that reforms must begin within the human mind, people struggle to overthrow oppressive political regimes, only to find themselves under the thumb of a new system as oppressive as the old one. Napoleon's dictatorship after the French Revolution and Stalin's rule after the Bolshevik Revolution are cases in point. The drive to subject others to one's will transforms man into the cruelest, most unfeeling of all beasts. Sacrificing one's self for the happiness of others transforms man and his world into something of great beauty. We must all realize this. And, realizing it, we must strive with our deepest, greatest power to bring about in the mind of every person on earth the inner – the human – revolution that turns man away from cruelty and suppression and in the direction of altruistic love and compassion.

Spiritual Values

IKEDA: Both the theory that man is innately good and the theory that he is innately evil have exerted strong influences on society. The former has in general stimulated liberalism, and the latter authoritarianism. The Judaeo-Christian idea that man first became sinful when he deliberately disobeyed God is, of course, set forth in the biblical account of Adam and Eve and their eating the fruit of the tree of knowledge of good and evil. Adam and Eve were totally good before they ate of the fruit but came to know evil afterwards. The Buddhist approach is different. We believe that life is potentially both good and evil and emphasize development of the capacity to discriminate between the two and act in the name of good. Making such judgements depends on the individual's spiritual values.

As everyone who has considered the matter realizes, Socrates was correct in saying that the important thing is not merely to live, but to live well. The selection of a spiritual attitude toward life that permits living well is the vital issue. What attitude toward life – and toward good and evil – best enables men to live well?

PECCEI: Spiritual values vary not only from culture to culture, but also from place to place; and over the course of time some Indian and Japanese Buddhist sects have interpreted the Buddha's teachings differently. I understand that such differences still persist among Buddhists just as they do among

87

Christian denominations. What is considered spiritually exemplary by the Sunnite Muslims of Saudi Arabia has something in common but need not to be identical with what is considered spiritual by their Moroccan brethren. As the centuries pass, spirituality evolves, adapting to changing situations. Spiritual values have evolved greatly in recent decades among the non-Catholic Christian laity and clergy, while the Catholic Church, which is more tradition-bound, has experienced difficulty controlling the most advanced members of its flock.

IKEDA: I agree that things like interpretations of the Buddha's teachings differ from time to time and from culture to culture, as the different views of Indian and Japanese Buddhist illustrate. However, in my opinion, spiritual values related to the fundamental meaning of good transcend time and culture. Furthermore, I believe that, on the basis of our shared understanding of the good, it is possible to finds means to enable human beings to understand and live in harmony with each other.

The most fundamental spiritual value is mutual help. We Buddhists strive to be enlightened to the spirit of compassion according to which good is respecting others and attempting to contribute to their happiness, and evil is sacrificing others to one's own selfish ends.

PECCEI: Very broadly speaking, higher human thought is, or ought to be, spiritually enlightened; that is, in constant communion with all aspects of our universal environment. Such thought should be permeated by a sense of intimate brotherhood among all the members of the vast family of man who are living now or have preceded and will succeed us everywhere. This oneness transcends bonds of individual family, race, creed, or nation and the kinships deriving from a common status or ideology. It should simultaneously be inspired by a feeling of community and belonging with nature, especially with all living creatures, no matter how different or distant from us. Spiritual values are often bred by a sense of our own ephemerality and pettiness in the face of the transcendent forces that keep everything in motion, and of superior beings that may exist but that human minds have neither encountered nor explained.

In other words, spirituality, which is latent or manifest in all of us, does or should lead us to strive for a creative and rewarding union with the entire Universe that we experience or may conceive. We may give the essence of the Universe a number of titles: God, or love, or even energy. Or we may not know what to call it. Still it gives, or should give, us supreme satisfaction to contribute our infinitesimal bit to its designs or continuity.

Individual Life, Greater Life

IKEDA: The Buddhist interpretation of life is one of an infinite entity manifest in repeated individualized entities – or human lives. Life and death are only different modes of the same greater entity. Though widely accepted in the past, this view is rejected by many people today in favor of the idea that one human life span is unique and complete within itself without being part of a greater life entity. Which view do you support?

PECCEI: Not even so enlightened a human being as the Buddha could answer all the questions that tantalize us. In all probability many of these questions either cannot be answered at all or are badly formulated by us humans. The Buddha conceived a fascinating vision of life in the Universe and transmitted it into a truth that can be believed but that, given man's present intellectual capacities, cannot be proved. Scientists generally posit what is called the theory of the Big

Bang, which supposedly occurred when a bulk of dense matter exploded upwards of ten billion years ago, and started the expansion process of the Universe that is apparently still under way. But this cannot be proved either. Therefore, in matters like the universal nature of life, I prefer to state my ignorance instead of affirming something beyond what I think I can understand. I am not a believer according to the definition provided by this or that faith and, therefore, share neither the Buddhist belief that my life will take another form after my bodily death nor the Christian belief in an immortal soul that, in an afterlife, will be punished or rewarded for deeds done during life. Nevertheless, I think I try in my small way to apply the Buddhist and Christian principle that – whether one is a believer or a nonbeliever – one should do good during life, and I try to do it without concern for either a rebirth or hope of reward or fear of punishment.

Although I believe that life on Earth will go on in one form or another, I am not sure that one can infer that human life itself will continue for ever and ever. To use a Buddhist concept somewhat lightly, I might say that our generations are building such a negative karmic account that, after they pass away, it may take quite a long time for human beings to appear on Earth again. In simpler terms, I believe in the continuity of life on our globe, not by means of a succession of individual rebirths and transmigrations, but by the evolution and interplay of all species. Since the human species is just one among a myriad of others – possibly not among the strongest or fittest existentially – it will be in greater danger if the planet's biophysical conditions undergo drastic changes. As a result, it may disappear from Earth. However, the Universe is so vast and the celestial bodies so innumerable that there is a very high probability – almost a certainty – that life as we know it, and possibly human-like life, too, may thrive on some planet similar to our own somewhere in the immense interstellar spaces. Frankly, however, I do not think there is anything more than this to say on the subject.

Now, I should like you kindly to clarify a few points of Buddhist doctrine on karma for me. First, at the time when the Buddha was living, the population of the whole world was only a few tens of millions; today, it is rapidly approaching five billion. If, through successive reincarnations, those few tens of millions have transmigrated in human form to our time, today there should be an equal number of people whose past lives have determined the mold of their present lives. What of all the other billions alive today for whom there have been no human predecessors on Earth? If they too had previous lives, we must logically infer either that human life certainly exists in parts of the universe other than Earth or that other forms of life have tended to ascend to humanhood. Is this interpretation correct?

To expand the question of karma still further, I wonder whether all those people who are richly endowed and powerful now were good enough in their past existences to merit their present privileged situations. Can the good fortune of a Hitler or a Stalin, both of whom enjoyed, but did not deserve, the best of material life and the privilege of power, be accounted for by very good deeds performed in previous existences? Why is it that the majority of citizens of the United States, generation after generation, enjoy a high standard of life and education and the best of health conditions while in Somaliland people are generally poor and hungry? Is it that individuals who have been especially good are grouped in one place, whereas those who have been relatively bad are grouped in another?

Second, the karma account carried over from previous lives affects each individual's present status; but at the same time we can improve our account and have those improvements inscribed in our personal record so that they can benefit us in future existences. I should like to know the range of conditions in our karma accounts that are permanently determined and those for which we have a certain freedom for improvement. The ability to improve one's karma seems to me to resemble what the

91

Christian tradition calls free will, another concept that I do not understand well, for God's omnipotence could – or should? – guide us to do only good. Please clarify the Buddhist concept for me.

IKEDA: I take your point. Buddhism teaches that life can manifest itself in many forms and that karma determines the form it will take. Being born a human being is the outcome of good karma. However a person – a particular manifestation of the general force called life – who ignores the value of being human and fails to live in a good, moral fashion may not be born human the next time. This could well account for the discrepancies in numbers of human beings in the distant past and now. In addition, the world of the Buddha is not limited to this planet but extends throughout the ten directions – north, south, east, west; north-east, south-east, south-west, north-west; and zenith and nadir – of the Universe, where there may be many other worlds inhabited by creatures at least as spiritually advanced as man. For this reason, an increase in the population of the world today over what it was a thousand years ago does not negate the eternal nature of life in general.

As to the question of whether all the rich and powerful of today earned their good fortune through good acts in the past, I think it is necessary to remember that wealth and power do not necessarily consitute happiness. Buddhist teaching holds that wealth of the body is better than wealth in the storehouse and that wealth of the mind is superior to wealth of the body. A man may be fortunate in having money and still be wretched in mental or physical health. His past karma is probably not very good. That of the man who, though poor in the wealth of the world, is healthy, trusted by other members of society, and happy in life is probably very good indeed.

A man's present state is not entirely determined by the immense, indefinite plexus of karmic causes inherited from the past. The efforts he has made after this present birth also play a determining part. Man enjoys more freedom than any of the other animals. The life of each individual contains various karmic elements, some definite, others only probable. Man is free; he has the choice of reacting to his definitive karmic elements in such a way as to produce either good or bad karma for the future. Take your own case as an illustration. You were karmically determined to be born at the time when the Fascists swept through Italy. You chose the way you would react to that situation. Instead of going along with the Fascists and producing bad karma for the future, you fought them to defend human dignity. In this way, you produced wonderful karma for the future. Hitler made the mistake of reacting to his condition in a way that can spell only the worst karmic results. The human revolution is a method of helping people to alter their manner of living, to react to their circumstances so as to produce good karmic results.

Helping Others Have and Use Liberty

IKEDA: As I have already said, the Buddhist law of cause and effect is not incompatible with free will. Although each human being inherits a karmic background from other existences, each is completely free to act in this world to alter it for better or worse, as he sees fit. In other words, in the Buddhist view man is innately free.

The freedom men have deserves maximum respect. Everyone should be free to deal with inherited destiny; but, in their greed for power, some human beings trample on the freedom of their fellows. We Buddhists believe that, no matter what advantages they reap by such actions now, sooner or later, retribution will come. However, karmic relations aside, the dignity and liberty of each human being deserve full respect. What concrete ways do

93

you think we should follow to demonstrate respect for man?

PECCEI: Respect for the freedom of other people to live as they wish is essential. It was vitally important in the time of the Buddha, when people were scattered in small, fairly dispersed communities, and is all the more important today, when great numbers of people are virtually piled on top of one another in sprawling cities, some of which – Calcutta, for instance – have a population probably larger than that of the whole of India in the Buddha's days.

The passive act of respecting the freedom of others is not enough. There are many people everywhere – near you, near me, near our readers – who, even when in principle free to choose their way of life, are prevented by socio-political or economic conditions from exerting their freedom. We must not only respect others' freedoms, but also endeavor to put all peoples in a position to exercise their freedoms. We can do this by contributing to an all-out, concerted effort to rid the world of mass ignorance, poverty, poor health, and other social ills, as well as of the political limitations on liberty that centralized power can apply now more easily than at any time in the past.

IKEDA: You prescribe social action, which I agree is of the utmost importance, yet, no matter how good the course of action, unless it is oriented primarily toward protecting the freedom of the individual, it can lead to bad instead of good. For instance, high-minded revolutions have often ushered in subsequent periods of rigid totalitarianism.

Democracy, Yes, No, or Maybe

IKEDA: Though they are difficult to put into effect in large societies, I am of the opinion that purely democratic means of

94

government must always be adhered to because diverging from them, even in times of crisis, produces tragic results. Obtaining a consensus and acting on it may take time, but it is crucial and, in the long run, efficient. In crises, autocracy may seem excusable because of the speed of action it permits, but I doubt whether this apparent efficiency can be either long-lived or fruitful. Do you advocate thorough-going adherence to democratic principles of government?

PECCEI: Yes, indeed. However we should not shut our eyes to the fact that, in the diversified and heterogeneous world of today, *different people interpret the essence of democracy differently*. For instance, the concept of democracy we have in Europe, America, or Japan is quite different from that of progressive democracy in the Soviet and other communist countries. And something quite different again is called democracy in various parts of the Third World. Even in the Western nations – including England, the mother of modern democracy – the democratic form of government, while the best form we have at present, is often hard put to be true to itself.

As currently organized, democracy was born when issues to be solved were relatively simple and could be taken directly or indirectly to the people. Today the problems societies face are so complex and difficult that even expert decision makers abreast of all aspects of human affairs cannot formulate clear ideas on most of them. No wonder, then, that public opinion is confused and that, when the occasion arises, voters either do not know how to cast their vote or can be swayed to vote as a party machine or pressure group tells them to do.

The democratic process, which has more than this one flaw, is easily circumvented even in countries professing its principles. Take the example of nuclear energy. Citizens with only inadequate knowledge of the technical, economic, and socio-political problems inherent in nuclear fission or fusion, of the deadly dangers of plutonium, and of the real availability and cost of alternative energy sources, are sometimes called to vote in referendums on nuclear power plants and may be inclined to vote yes this year and no in another referendum on the same question in two or three years' time. Obviously, plans for nuclear generators cannot be adopted then dropped every year or so. Moreover, not only the construction, but also the operation and dismantling of these power plants, involves so great an outlay of effort and money, and demands so many safety precautions for such long periods of time, that all decisions made today will inevitably have great implications tomorrow. Choices made now, often in connection with short-term interests or expectations by a generally not fully informed citizenry, will in effect impose costs and risks that are difficult to calculate on future generations – whether the plants are built with present-day technology or the nuclear programs are stopped without the development of other alternative energy sources. How can this dilemma be faced equitably in a truly democratic spirit?

The situation was and is further aggravated by uncertainties and emotions manipulated by certain well-organized pressure groups. Under these conditions, governments wanting to institute nuclear-energy programs have opted to follow the quite undemocratic shortcut of going ahead without proper popular consultation. A case in point is that of undoubtedly democratic France. Faced by this kind of confused situation, former President Giscard d'Estaing maintained that, if the plan – indispensable in his view – to build a large number of breeding reactors, which are currently considered potentially very dangerous, had been submitted to parliament or had been put to referendum, it would certainly have been stalemated. His government therefore took sole responsibility for launching the largest nuclear program in the West.

An example in another field is that of war. Some 150 wars

96

have been waged since World War 11, and almost none has been declared after due democratic process. At present, the two superpowers are actively preparing for nuclear warfare – they claim they will fight on defensive grounds to deter an adversary. The essence of their preparations and their military strategy is, however, classified, making popular consultation technically impossible. Should the so-called enemy attack with nuclear missiles, there would be perhaps five or six minutes to retaliate; the decision would be made by a score of officials – or a computer – not by the sovereign people.

The principles of, and conditions for, democratic governance of contemporary society must be thoroughly rethought and reformulated. Fundamentally, of course, the people must remain the arbiters on major issues concerning their lives and futures. Their opinions should dictate the policies and strategies to be adopted and implemented. To make this possible, however, not only the people themselves, but also the political class must be better educated and prepared; and the mechanisms of the democratic process must be substantially improved. Such evolution will, of course, require great political acumen and good-will, as well as time, yet it cannot sufficiently change the situation by itself. Our institutions were designed in and for other times and must be reformed, at both the national and international levels. More important still, our ways of thinking and behaving must be updated to face the new range of challenges all of us must meet, individually and collectively, within our own countries and, transcending all boundaries, through the whole vast world.

In short, though I fully endorse the principles of democracy, the answer to your question cannot be a simple yes, some clarifications must accompany it. The question of what kind of democracy is desirable and possible in the years and decades to come in individual countries, or in the global community, does not allow for a simple answer. As in the case of other major

97

issues confronting us at this complex turning point in history, we must consider a wider context of problems and alternative solutions before deciding what position to take.

IKEDA: I agree with you that this is too complicated an issue for a clear-cut yes or no. As you suggest, today many issues that cannot be correctly decided without great specialized knowledge are entrusted to political organizations. Nonetheless, it is fearsome to imagine what might happen if, because there are many problems of this kind, democratic processes were to be ignored and the fate of mankind decided on an autocratic basis.

Since many issues are beyond the ken of the non-specialist, a nation could in theory be militarized and sent to war without any explanation having been offered to the citizens and without consultations having been taken with representative governing assemblies. I am firmly convinced that governments must be required to make the majority of the people aware of the problems concerning the destiny of mankind and that all judgments and decisions made on such problems must be made in a democratic manner.

Peccei; You are right. It still remains to be seen on what bases and by what ways and means this can be made to happen in future years.

Peace a State of the Spirit

IKEDA: Many of the crises facing the world today threaten man with annihilation, but none in so cruel and horrifying a fashion as war. Buddhist philosophy makes it possible to interpret death in battle as karmically determined. As I have said, the will of the living person can change karma and not everything is determined by causes from the past. Choosing to

98

fight and kill now undoubtedly will have bad karmic effects in the future.

In modern wars, people who have no will to fight and who are unprepared to defend themselves are killed and wounded – for instance, millions of noncombatants were killed and wounded in the atomic bombings at the end of World War 11.

Mutual slaughter among soldiers is cruel and has very bad karmic effects, but the evil effects of slaughtering unprotected ordinary citizens are graver by far. The knights – and samurai – of the past protected themselves with armor. The combatant soldier of today, though armorless, still has means of protecting himself and is trained in their use. Ordinary people lack both the means and the training and are killed indiscriminately. My discussions with American, Chinese, and Russian leaders in many fields are based on the hope of averting the catastrophe of war, the most destructive and wasteful of all human activities. The most reliable basis for world peace is to strengthen bonds of mutual trust and understanding among all peoples. The place to start achieving this end is the United Nations. Do you consider the establishment of a global government a possibility? And, if it is possible, could such a government lead to peace?

PECCEI: I do not think that the United Nations in its present configuration can either function as the cornerstone of a world political organization that provides an adequate response to the exigencies of our time, or be the ideal meeting place of a multibillion population, diverse in standards of life and imbued with a nationalistic, beggar-thy-neighbor spirit. Yet, as we have no other international forum of comparable importance and almost global representativity, we must keep it and strengthen it, no matter how desultory its performance in general and its peace-keeping record in particular. Though humankind has radically transformed its own condition on the planet and enormously improved its material equipment, it has failed to liberate itself from antiquated political philosophies and institutions. This, however, is not the fault of the United Nations, which has become the virtual repository of all national rivalries and international tensions, insufficiencies and weaknesses. Under such conditions, the United Nations can do very little to reestablish order in the world or even intervene authoritatively and settle disputes that may endanger peace. For more than thirty years, the United Nations has proved ineffective when the chips were really down. To consider only recent events, it has been unable to do anything to prevent endemic warfare between Israelis and Arabs, to stop fighting in the Horn of Africa, to bring the Iran-Iraq war to a halt, to settle the Falkland-Malvinas dispute, or to avoid the holocaust in Lebanon. The Second United Nations Special Session on Disarmament, held in May 1982, was a pathetic and dramatic failure.

All this strongly suggests that, though the United Nations organization may be a fairly useful tool for diplomacy and debate at government level, it has no practical power or influence on the now crucially important issue of war and peace. The United Nations and the world polity it subsumes must be profoundly reformed if they are to become promoters or guardians of peace.

You also ask whether a global government would be a good way to bring peace. I do not think so. For the moment, the constitution of such a supergovernment is quite unthinkable anyway, and it would be depreciative from a number of viewpoints, not the least of which is the difficulty of organizing and making it function democratically. We should not pursue the chimera of a central world government at all costs. In this, as in many other instances, we should take a hint from the world of nature. No one central organ or system controls a living body or the evolution of a forest or an ocean or presides over relations among the various natural environments of a valley. A multitude of large and small systems exist and interlock in all these biological organisms and biophysical complexes, and dynamic equilibria are maintained among them by mutual checks and balances. These are the ways of life. Being a self-organizing living system, human society should pattern itself on these other evolving systems. No illusion of finding a short-cut to peace should induce us to pursue the path of a world government.

IKEDA: I agree. When the late Arnold J. Toynbee and I discussed this same issue, I said that, though I feel a global governing body will become necessary in the future, I think it must be federalist, and not totalitarian, in nature. Taking Chinese history as his example, Mr Toynbee said that in the first stage a powerful state establishes unity and that in the second stage this is succeeded by a comparatively moderate dispensation. He was expressing no approbation of this kind of evolution but merely said that, as long as human nature remains unchanged, such a course of development is probably inevitable.[1]

My own opinion was and remains rather different. I think the human sacrifice demanded by the creation of a strong state is too great to tolerate. To avert it, I think a change in human nature is needed; I feel that a federal form of world government could become the basis for the required change.

You seem more pessimistic than I am on this topic. Still, if I interpret your meaning correctly, you do not reject the importance of a world government outright but think that one could be formed as a place of agreement if human consciousness were to attain a higher plane through natural means.

[1] *Choose Life – A Dialogue,* Oxford University Press, London, 1976, & *The Toynbee-Ikeda Dialogue, Man Himself Must Choose,* Kodansha International, Tokyo, 1976. (In Spanish, *Escoge la Vida,* Emece Editores, Buenos Aires, 1980: In French, *Choisis la vie,* Albin Michel, Paris, 1981: In German, *Wähle das Leben!,* claassen, Düsseldorf, 1982).

Since a whole range of other problems, including natural resources, energy, environmental pollution, food supplies, epidemics, and information now demand cooperation on a worldwide scale, I feel that global government must come into being as a place in which to deal with these issues. Of course, the most important matter such a government would have to deal with is peace.

PECCEI: The question of peace is so fundamental for the survival of our species that I should like to make several other points. To start with, *a clear distinction must be drawn between security, disarmament, and peace.* Security is, of course, a legitimate aspiration for every person, community and society. However not even the disorder and commotion of our epoch can explain why our generations are fainthearted and muddleheaded enough to entertain the grotesque belief that security can be bought by piling up mass-destruction armaments that constantly grow more powerful. Yet, for quite a few years public opinion in the most advanced countries of both East and West seemed to have believed that security could indeed be attained by relying on the ultimate weapons, while governments were and are doing their utmost to convince their citizens that this is the right way and that the sacrifices it involves are inevitable.

Fortunately, it is now dawning on quite a few people that the armament frenzy of some of our ruling classes and its perverse, Kafkaesque rationale are merely syndromes of the general state of malaise and disarray pervading modern societies, and that the imperative is to restore a sound state of social health. A real turning point, however, will be reached only when enough citizens are ready to stand up and declare that the issue of security is being altogether wrongly addressed by strengthening the infernal war machine and that, while we may like to live under the illusion that we are protected from the so-called enemy

102

in this way, we are actually courting disaster for one and all.

Recognizing this fatal mistake will, at the same time, show how foolish it is to equate disarmament with peace, or to believe that the war machine can actually be dismantled piece by piece by disarmament negotiations carried out by exclusive military and diplomatic circles under the cloak of secrecy. So far this delusion has remained unassailed. Since the end of World War 11, such efforts have continued, almost without interruption, with the aim of promoting measures and devices to prevent or resolve conflicts, of reaching a modicum of arms control and limitation, of establishing nuclear-free zones, of making possible mutual inspection of military build-ups and deployments, and of fostering general disarmament. All the while, however, the destructive capacity of the nuclear arsenals has increased more than a millionfold.

Although the campaign for disarmament must not slacken, we must face the hard fact that up to now it has been a monumental failure. And, what is more, while elaborate negotiations are under way at the time of writing, new and terrific laser, particle-beam, chemical, biological and meteorological weapons are in the military pipeline chiefly for use in a space war, or a star war, that would be the last word in the ultimate folly of our obtuse belligerence. All the sciences and disciplines are being marshaled to prepare for this new kind of warfare, colossal means are being mobilized, and an unprecedented propaganda machine – capable of obliterating in a moment all the advances made on paper at the negotiating table – is being readied.

Let us realize that, in any case, even a *bona fide* disarmament agreement would be only the first step in the right direction, for disarmament is not peace. Disarmament can be a key factor in achieving peace, a milestone on the road to peace, but it is not peace itself. *Peace is an intangible value, a cultural state of soul and mind,* that must be so clear and strong within each individual and so widely hailed as a vital necessity by other people that it becomes the common patrimony of society at large. Peace will thus come to exist only when all, or a wide majority, of the citizens come to treasure it as something precious and worthy of commitment. Whereas war is the distilled gall of arrogance, egoism, mutual distrust and fear, and is almost invariably brewed up by the wielders of power, peace is the natural outcome of mutual comprehension, tolerance, respect, and solidarity among people – and can spring only from the heart of the people themselves. As for disarmament, it lies in the limbo between war and peace and is decided upon by only a few people. We must convince ourselves that disarmament, however essential, cannot be enough in itself. Any relief of tensions it could bring about or any betterment of political climate and economic situations it could create would be inherently precarious, too easily upset by a change of moods by warlords. No true peace can be established on Earth unless and until the so-called military-political-industrial-scientific warmongering complex is eradicated, and human society as a whole really becomes peace-loving and peaceful.

You are the head of a highly important peace-oriented Buddhist sect. What do you say about the inane efforts made so far to bring about disarmament in the world? How can the cause of peace be furthered? Is there any forceful action that can be undertaken by ordinary men and women in various parts of the world to let the reason of peace win over the unreason of war?

IKEDA: I can answer your questions by referring to what we Buddhists consider the more general causes of all human inanity and unreason. In Buddhist teachings, the three spiritual defilements called the Three Poisons of greed, wrath, and folly are the fundamental causes of the Three Calamities of war, pestilence, and famine. To eliminate the three calamities it is necessary to eliminate their fundamental causes: the Buddhist teachings set forth a way of doing precisely this. That way manifests the great wisdom and immense compassion that the human spirit is endowed with and thereby opens the way to triumph over greed, wrath, and folly and to the establishment of peace in the form of what you call a cutural state of soul and mind.

Allowing the Three Poisons to run rampant and uncontrolled is a state of nonculture. Employing wisdom and compassion to bring them under control, as one cultivates a field, tames a wild horse, makes constructive use of potentially dangerous fire, or uses a toxic substance to effect a cure, is a state of culture. In the past, human beings have concentrated effort on taming external materials and forces, while allowing the untamed forces within their own inner selves to run wild. Though ethics and morality have made some attempts in the right direction, they have never delved into the vast power that is deep within the mind. Attempts at disarmament have failed because they have dealt with superficial things and have not faced the need for a fundamental inner revolution in all human beings. The sponsoring of such a revolution on the basis of Buddhist teachings seems to me to be the most important action ordinary men and women can take for the sake of peace.

Outmoded National States

IKEDA: The system of national states served the historical role of reducing friction among still smaller subdivisions of human society by organizing them into cooperative bodies. Today, in spite of the crying need for cooperative relations on a scale bigger than that of the national state, opposition and conflict persist and are even increasing.

Japan is an example of the process whereby unification of a single state can result in a cooperative system. Before the early seventeenth century, Japan was ruled by various feudal lords who squabbled among themselves. In the fifteenth and sixteenth centuries, the conflicts were resolved. Peace and unification of the nation under one government were achieved in the seventeenth century, but a nationwide system of truly mutual cooperation was not forthcoming until the second half of the nineteenth century, when everything was organized for the sake of modernization and Westernization.

Like the Japan of the past, today many national states have established domestic unity and harmony. However, under the existing system, as national states they contribute to the aggravation of conflict and opposition in international society. One of the most pressing issues facing us today is the creation of an international society that will foster mutually cooperative relations among all human beings. What is your opinion of abandoning the present system of national states, who protect themselves or further their interests by means of military force, and the creation of a larger body outside the national state framework for the solution of worldwide problems like conservation of natural resources, environment pollution, food shortages, and so on?

PECCEI: Yes, the nation-state system definitely needs profound revision – as soon as is safely feasible. The origin of the modern nation-state can be traced back to the Peace of Westphalia of 1648, which put an end to feudalism and was a socio-political milestone in the history of Europe. Our continent was then sparsely populated and covered with forests; those few who could read did so by flickering candlelight; and everyone traveled by foot, except for a very few people who rode horseback and even fewer who could afford utterly uncomfortable carriages. Today, in spite of our globe-girdling communi-

106

cation and transport systems, in spite of the almost unbelievable development of knowledge and instant information and of computers and satellites, and in spite of The Bomb, we still cling to the nation-state as the backbone of the world's political organization. This outdated model, coupled with the divisive and conflict-fostering principle of national sovereignty, is one of the major roadblocks hampering humanity's progress in this otherwise impetuous age of change.

Structural and philosophical evolution is indispensable. One of the first requirements is to *denationalize the world community*. I see four major ways to achieve this. First, the development of *subnational* autonomies and authorities. Large private enterprises do not always provide a good example of public spirit but, organization-wise, they set the pace. Their rule is to decentralize decisions and let them be made at the hierarchical level where they will be implemented, instituting ways and means to harmonize and coordinate the whole. Similar arrangements should be devised and adapted to the political and administrative organization. This is a case in which the new fashionable formula 'Think globally, act locally' can be better applied. The principle of decentralizing political and administrative functions and decision making, putting them in the hands of people who are directly involved and better aware of problems and possibilities, has already undergone felicitous or experimental applications both in North America and Western Europe.

IKEDA: In the case of private business, as such firms as International Business Machines and Coca-Cola show, international expansion can be successful. Though organizational systems vary, in cases of this kind, it is usual for the home offices to provide technological guidance but to leave personnel and labor matter and production goals up to the local organization to the maximum extent possible. In quite a few unsuccessful cases, however, intent on maintaining their own authority, out of shortsightedness and narrow-mindedness, home organizations meddle in affairs that should be left to local people and thus reduce incentive. Of course, realizing that poor performances can mean business failure, wise private concerns immediately take steps to rectify this situation. In this sense, you are quite right to say that large private enterprises set the pace, organization-wise.

The case with political and administrative organizations could be quite different if such groups – or other authoritarian organizations like religious denominations – attempt to force all decentralized decision-making bodies to function exactly as the home organization functions. Trying to put everything under home control will prevent decentralized organizations from taking firm root in the minds of the local people. In this respect, I am interested in the idea of federation as opposed to decentralization.

PECCEI: A second way of denationalizing the world system is the evolution of *supranational* federations, unions or communities of nations that share certain interests and should logically create joint institutions, delegating to them a number of economic, political and other functions. The best example so far, although its consolidation occurs at a snail's pace, is the European Community. After two world wars originated and fought mainly in Europe, it would have been unconceivable for West Germany, France, Italy, the United Kingdom, and so on to try to go it alone at the risk of yet another war. At the initiative of a few leading statesmen and with the support of public opinion, an founding group of European nations decided it was time to stop utilizing individual mechanisms and adopting independent policies on matters which they could only approach at a higher level of concern to the whole of Europe. The seeds of the European Community were cast.

In 1980, The Club of Rome sponsored a conference at United Nations headquarters to examine the possibility and practicality of simplifying and making more effective dialogue and relations among the world nation-states, by promoting the formation of regional unions among countries having a common cultural tradition or sharing the same geo-economic area. Such regional groups, which would concert among themselves all major policies, might provide a middle road solution between the current fractious, chaotic, ungovernable set up, in which one hundred and sixty egocentric, often unviable, sovereign states are engaged in unending squabbles, and the world government, the very idea of which is destined to remain as unreal and misleading as a mirage for a long time to come. A small unit has been established in the United Nations, to study likely programmes for regional and interregional cooperation in the 1980s, opening the way for world cooperation later on.

IKEDA: The idea of forming regional unions among countries with a common cultural tradition or the same geo-economic sphere is outstanding. Of course, peace demands global cooperation, but there are many problems directly related to the ordinary daily lives of the people as a whole that require cooperation on a plane transcending national borders. The cultural common ground between the European nations is much greater than the differences which exist among them. Furthermore, in economic terms they are very closely bound together. Mere geographic proximity has meant that national life styles have mutually influenced each other because of the possibility of frequent contact. Daily necessity has bound the European peoples into a kind of internationalism that I feel is difficult to destroy. When war breaks out, the vital importance of cross-border contacts becomes painfully apparent.

I believe that internationalism born of frequent contact on matters of daily necessity can break down barriers and make the national state an outmoded relic of the past. If the kind of process taking place in the European Community, no matter how slowly, could expand to link East and West, the United States and the Soviet Union, Japan, China, and the Soviet Union – a power for peace too great to be ignored would emerge. In other words, transnational interests sponsor mutual trust on the basis of mutual need.

PECCEI: Yes, and today *transnational* interests and organizations provide a third method of bypassing the fragmentation of the world into national enclaves. Large corporations were the first to perceive the advantage of thinking, organizing and operating across frontiers. Whatever strong-arm and cavalier attitudes they have sometimes entertained, they are more in tune with certain realities of our times than stubborn advocates of rigid compartmentalization of the world into sovereign national enclaves. The world community must study how local, national and regional prerogatives can be satisfied without renouncing transnationalization of the human enterprise in our ever more integrated and interdependent global system.

Fourthly, we must liberate our minds from the current, all-pervasive, obsessive reference to nationality. Most of the things that make for a high quality of life are *non-national* and do not recognize and are not constrained by artificial human frontiers: music, the arts, knowledge, love and compassion among others. The sun, the winds, ubiquitous nature and, on the other side of the coin, pollutions, acid rains, and the consequences of the depletion of the upper-atmosphere ozone layer, exist or occur across boundaries. It will be a happy day when we finally come to realize that both the concept and practice of dividing the world into an assortment of inward-looking nation-states prevent us from making good use of our planet and multiply the detrimental effects of the bad things we usually do.

Finally, each of us is multivalent. It is absurd to ignore this and to force people into one or another suffocating national mold. Modern men and women are more than simply the citizens of a certain national state. For instance, by birth and culture, I am a Roman, an Italian, a European, and a Westerner and, by election, a world citizen eager to understand and live with and for all my brethren – whatever passports they may carry.

IKEDA: I agree entirely, but all too frequently people, while knowing that they are, for example, a Japanese, an Asian and a member of the human race, give precedence to a narrower concept of self. Such constricted views have led the people of my nation to go to war with the peoples of the United States and China in spite of a realization of the ugliness and folly of human beings' killing one another. On a less disastrous level, there are many people who give precedence to their own profit or to that of their families over the welfare and happiness of human society as a whole or who, as you often point out, are more concerned about the short-term benefits they may enjoy than about the well-being of future generations. Buddhism defines this kind of selfishness as evil and calls contributions to the welfare of others and society as a whole good. For the sake of doing this kind of good, we must all give precedence to our multivalence on the widest possible scale – and this means believing firmly that we are all brothers.

PECCEI: Let me add a word about *the principle of sovereignty,* which has become closely linked with the concept of national statehood. I have demonstrated elsewhere[1] how this principle lies at the root of many of the evils, dysfunctions, and dangers

[1] *The Human Quality*, Pergamon Press, Oxford, 1977. (In German, *Die Qualität des Menschen*, DVA, Hamburg, 1977); *One Hundred Pages for the Future*, Pergamon Press, Oxford, 1981 and Futura, London, 1982. (In German: *Die Zunkunft in unserer Hand*, Molden-Seewald, Munich, 1981).

inherent in the present *inter-national* organization of human society. Today, it is customary to speak of macroproblems, megatrends, planetary challenges and global alternatives. No matter how exaggerated these definitions of the new dimensions of the enlarged environment (affecting, perhaps decisively, the lives of all of us and therefore demanding our continuous attention) may be, it is certain that the dominant realities of our time cannot be broken down to suit the artificial patterns of our national boundaries. If we want to move ahead into the future with a sporting chance of success and survival, we must purge and purify our minds of the myth of sovereignty, which is a political and philosophical leftover from a dead past.

More Communications, Less Communion

IKEDA: Advances in communication technology now make it possible not only to talk with someone on the other side of the globe but also, through television, to see his facial expressions. Communications and mental interaction have not necessarily developed to parallel the increasing sophistication of physical communication. I believe that the peoples of the ancient world probably communicated with each other more often and more intimately than do the peoples of the current period of nationalism.

The ancient Silk Road was an artery along which passed, in addition of course to the fabric for which it was named, arts, religion, and learning. In a physical and spiritual way it linked not only the cities of China and the Middle East but also, through its various ramifications, the nations of the Far East and Italy, France, and Spain in the West. It was along this road that Buddhism passed from India into China. From China, by way of Korea, it then passed on to Japan. The many Chinese and Korean immigrants who traveled there, immersed in the spirit of Buddhism, greatly enriched the culture of Japan by introducing it to the civilization of the Asian continent.

At that time people were not hindered by the complicated immigration and naturalization processes that have been established by modern national states. An alien who got along well with the people of his new country could easily become a member of its society. In spite of advances in transportation and communications, today people are hindered in efforts to establish true person-to-person cross-cultural contacts and exchanges by the red tape of the modern national-state system. I should like to see this red tape cut away as quickly as possible, because I believe that understanding based on contact and knowledge of the ways of living and thinking of other peoples is one key to averting catastrophe and building a bright future.

PECCEI: I disagree with your view in one respect and agree in another. With modern media, telephones, television, videotapes, cables, Telex, and airmail and with other kinds of electronic devices already in the pipeline, interpersonal communication is vastly greater now than ever before. Every day, if not every hour, using one method or another, we can easily get in touch with many people, both near and far.

However *much of our communication lacks communion*, the personal touch, the warmth of the human presence and of mutual knowledge. And this is where I fully agree with you. Communication of this kind is electronic and artificial, involving little face-to-face, heart-to-heart contact. Passions are quenched; even the invective loses its sting. I can call anyone I like on the telephone, but I can hear only his or her mechanically reproduced voice. In television, the one-way interchange is prepared by a few people for the many, who merely look and listen. Conferences may be held on telescreens without the conferees coming together, shaking hands or patting each other's shoulders. In other fields, people are becoming so accustomed to talking to trusted machines and getting quick, pertinent answers from them that they no longer find need or pleasure in talking to other individuals, who may fumble or disagree. You are right. The consequence is that well organized pressure groups can influence large numbers of people who walk side by side but remain alien. Because the crowd is generally amorphous, the city anonymous, and the family – the very heart of society – crippled, all are easy prey to those who seek power.

In Japan and other frontline technological nations, vanguards of busy researchers and entrepreneurs are building a still higher threshold of progress: the information society. Although we are full of admiration for it, a few puzzling questions come to mind. How many people will be able to adjust their lives to the new thresholds? Will ordinary citizens be induced to communicate still more and commune still less? Or is it possible to make this kind of progress happen not as something extraneous to and artificially superimposed on the billions of the world's ordinary citizens, but as something as easily acquirable as an extension of innate capacities, similar to playing music, climbing a mountain, conceiving God, crossing a traffic-jammed thoroughfare, or progressively learning one's own language or dialect?

Tolerance for the Joy of It

IKEDA: In spite of its wide diffusion among different peoples – witness the Nazis Germans' hatred of the Jews or the unfair treatment accorded to black people by many citizens of the United States – I do not agree with those who claim racial prejudices to be instinctive in man. White, black, yellow, or red children relate with each other openly and without prejudice. Hostility only develops later, when children grow up and identify with different racial groups. Do you believe that cultural traditions and social customs play a major role in racial prejudices?

PECCEI: Yes, I do. And to your category of racial prejudices, I would add religious, ideological, and national prejudices, not to mention prejudices related to sex and age. As you point out, children get along well together, no matter what their color and family position. This is because child play-culture is similar everywhere. As children grow older and become part of adult culture, however, differences assume importance; as a result color lines, sex discrimination, status bias, class consciousness, and so on appear and cause trouble. The main source of friction between white and black peoples or between the older immigrants and the new Hispanic groups in the United States, the ground for the bloody strife between Catholics and Protestants in Ulster, and the often hidden motives for the persecution of minorities taking place at an appalling pace in many parts of the world are all basically cultural. Not that different cultures are in themselves antagonistic; to a large extent, the fault lies with the ruling elites, who find it expedient to exacerbate cultural differences as leverage to maintain their power.

It is all the more lamentable that such differences can become instrumental in generating conflict because *cultural diversity is highly enriching.* Imagine the horror of the advent of one worldwide, standard culture – which is, however, coming into being and is jokingly called the world Coca-Cola-nization. Gone would be the charm, stimuli, and uplift of learning derived from other peoples' cultural backgrounds, different outlooks, values, and ways of thinking.

IKEDA: You are absolutely right. The popularity of travel shows that human beings are curious. We can all learn from different things and different places. Acquiring fresh knowledge and becoming acquainted with unfamiliar sights and ideas constitute the joy of traveling. A person who is afraid of the unknown, antagonistic toward the unfamiliar, and stuck in his old familiar rut is already senile, no matter what his age.

PECCEI: Yes, of course. I am delighted that you are Japanese and Buddhist, not Italian and imbued with the same traditions as I am, because I can learn a great deal from you – more than from my fellow countrymen. I hope that you can learn something new from me, too. The tolerance and understanding that can result from the intermingling of cultures is more significant than personal interest and gain. I like to say to certain friends of mine: "So, you are a Marxist. Good. Teach me something about your insights and rationale." Or I tell my African colleagues: "Explain your traditional wisdom and experience to me. Let's share. We'll both be richer for it." The development of mutual comprehension and tolerance, as well as eager acceptance of cultural diversity instead of the kind of cultural self-righteousness and distrust of outsiders that we currently exalt, will create a new atmosphere in the world and make our common *problematique* easier to tackle.

Willy-nilly, all of us members of many races and cultures must live together on this planet. We would be much happier and safer if we learned to understand and share for the joy of it – instead of grudgingly putting up with each other just because there is no other way of getting along.

The Role of Religion: Is Ecumenism Possible?

PECCEI The human psyche is being affected – although we do not yet know exactly how – by new factors and phenomena alien to the experiences of our predecessors. Let me just mention a few of these new aspects of contemporary life. On the same old planet we now crowd many times more people than was even imaginable in the past. Moreover, individual mobility is far greater than the kind our ancestors attributed to certain

116

magicians and we can communicate among ourselves worldwide more intensely – with the help of special gadgets – than was once possible among members of a single clan living together. At the same time, our power to transform our habitat beyond recognition has grown immeasurably; we can destroy millions or billions of our fellow humans by giving orders to other gadgets accurately prepared for this purpose. All these changes cross-influence each other continuously, with the result that human pressure and impact on everything on Earth is thousands of times greater than it used to be. The natural systems and cycles – particularly the biological ones – are profoundly affected by these changes, as I have tried to show before. So is the human psyche. Our reactions are mixed and still very confused: disarray, a sense of helplessness, alienation, loss of faith, fear of other people, recourse to violence, wild hope in technological miracles, or yearning for charismatic leadership. *Have the different religions risen to the challenge of this epochal crisis?* To what extent are the different faiths and creeds really helping poor modern man in his predicament? Have the various Buddhist sects united in a common effort to foster the good of all? Have the Christian denominations shelved doctrinal differences to bring hope to all peoples and enlighten the conscience of the individual everywhere, whatever beliefs he or she holds? Are the competing families of Islam ready to forget for a while their feuds and embrace, without distinction, not only all the faithful, but other believers or nonbelievers too, and accompany them in the quest of truth and understanding? Have all these great religions ever called on the other major faiths to work and learn together how to guide the world population away from its present plight and toward terrestrial salvation before it is too late? Or are all the churches, faiths, and religions destined to go it alone forever, each one proclaiming its own creed and minding its own business as usual, as if the world had not changed? Though diverging not so much on the essence of

117

the good as on its revelation, formulation, or interpretation, will religions remain incapable of converging in a superior vision of what combining their unparalleled spiritual and moral forces could do to make humanity better and thus to open roads to a future reality worth living?

Is ecumenism just a figure of rhetoric? Will doctrinal sovereignty in matters of faith prevent ecumenic synergies from becoming something to be counted on in the real world? Or can the great inspiration of ecumenism be brought to the practical expression of religious accord, solidarity and cooperation in the pursuance of the good of all men? What is your opinion? What initiative is Soka Gakkai willing to take in this connection?

IKEDA: The questions you ask are stern ones, demanding honest and serious thought. To be perfectly frank, I consider the ecumenical movements being sponsored by a number of religions today nothing short of deception. Any religion, to remain a religion, must be convinced that its own doctrines are uniquely correct and that other doctrines are mistaken. In actual practice, ecumenism would demand a compromise on doctrinal matters that I consider unacceptable.

This is not to say, however, that religions are to stand by and allow modern man to sink still deeper in perilous confusion. As you have remarked, races of different cultural backgrounds must coexist on the planet. I believe that cooperation among religious bodies, though impossible on the plane of doctrine and teaching, is not only possible, but also essential, on the quite different planes of politics, economy, industry, and culture. In other words, though I think cooperation in terms of religious doctrine is tantamount to the rejection of religion itself, I feel that peoples of various religious faiths ought to be encouraged to cooperate on the political, educational and cultural levels.

It is the pollutants in human life – in Buddhist terms they are called The Three Poisons: greed, anger, and folly – that stimulate man to prejudices and hatred and breed confusion and unprincipled conflict. In other words, the causes of all unhappiness are to be found within human life itself. Christianity, Islam and Buddhism have taught love and compassion transcending differences of race and ideology as the ways to promote mutual understanding and to bring about a kind of ecumenical union of mankind. Yet love and compassion alone have been unable to eradicate the greed, anger and folly that are at the heart of the trouble. Furthermore, movements for the ecumenical unity of mankind based on such ideas have either never progressed or, worse, have tended to foster hatred and contempt among believers of different faiths. Insignificant dogmatic or ceremonial differences have frequently split the members of one religion into factions more inimical to each other than to adherents of totally different creeds. In addition these differences have also sometimes goaded people into placing less value on the lives of fellow human beings than they do on the lives of non-human animals.

The blame for this kind of hatred cannot, of course, be put on the spirit of love and compassion but must fall on the greed, hatred and folly that are inherent in life. Religion is the force that can control these undesirable human traits. The undeniable truth that they have not always been controlled seems to point to the impotence of religion. This is not necessarily accurate. The value of a religion depends in its ability to supress greed, hatred and folly, and in this way to allow the spirit of love and compassion to emerge. (This is why I embrace a religion that teaches a human revolution of the inner being.) In other words, a religious teaching is incapable of suppressing the dark side of human nature, but must be the means whereby the individual builds within himself the strength that, combined with his own willpower, can overcome the Three Poisons.

I am, therefore, confident that religion, especially the higher religions, must come to play a bigger role in the minds of men

119

because, without them, humanity lacks a way to combat greed, hatred and folly, and to follow a path leading to mutual understanding and respect. Only lofty religious belief can help us achieve global peace and retrieve mankind from the brink of the precipice of nuclear war and annihilation.

As I said earlier, however, religions should not compromise on points of fundamental doctrine but should boldy assert their profound truths. Nonetheless, there must never be physical struggle among religions. Should verbal dispute arise, when the debate is over, the party that has been bettered should frankly and good-naturedly admit it.

In brief, it is my belief that if each individual believer in each of the high-level religions sincerely and diligently put the teachings of his faith into practice, it would be possible to avert the crises facing mankind today, while at the same time preserving intact the doctrinal individuality and independence of all faiths.

PART III   THE HUMAN REVOLUTION
Views of Daisaku Ikeda
and
Views of Aurelio Peccei

THE HUMAN REVOLUTION

Views of Daisaku Ikeda

Human civilization has concentrated on altering environmental conditions to suit man's needs and desires. Science and technology have employed natural powers to develop to the point where they can overcome nature itself. In addition, man has striven to improve legal and social systems in the belief that, with such things perfected and with science and technology raised to a high level of sophistication, humankind will find ideal, long-lasting happiness. Scientists, technologists, politicians and legal experts have taken pride in their work because they believe that, in carrying it out, they were bettering the condition of mankind as a whole. In turn, society has valued and respected these people.

Their glory and the trust it inspired have now come under a cloud of doubt. Many people have come to suspect that the very specialists who contributed most to the advancement of human culture are leading manking to destruction. Many Americans – and some young groups in Japan too – have rejected the benefits of science and modern living. Though not espousing the totally primitive, they have moved to the wilds where they have formed communes, living in contact and harmony with the world of nature. They are relatively few in number, and their actions are extreme. Nevertheless, a considerable segment of the ordinary citizenry, while refraining for various reasons from following suit, sympathize with these extremists.

A transition has taken place in general attitudes. The

apologists who brimmed over with hope for the future a little over a decade ago first grew cautious then became openly pessimistic. Formerly only highly sensitive people – poets, artists and visionaries – saw inhumanity in scientific technology and took a dim view of man's future. Today, scientists them-selves point to the fearsome pitfalls in store for mankind.

In the field of legal and social systems, too, things have turned out less well than was hoped. The extent to which the weak and underprivileged can enjoy a secure way of life is an important gauge of the progressiveness of a society. It is good that wel-fare systems have been worked out to guarantee a minimum standard of living to the unemployed, the hanidcapped, the elderly, and so on. However the solution of material difficulties through well-established welfare systems does not necessarily save people from spiritual and mental suffering. Indeed, often the more materially blessed the society, the higher the level of mental insecurity and dissatisfaction. The number of mental patients was small in Japan during the materially difficult years of World War 11 and the immediate postwar period. It radically increased as the nation became prosperous.

Inherent in the pursuit of material prosperity itself are such evils as environmental pollution and exhaution of natural resources. Perhaps mankind will solve these difficulties by means of science and technology. Solar energy might be used, materials might be recycled, perhaps indefinitely. Nevertheless man can no longer draw irresponsibly optimistic conclusions about the future on the basis of science and technology alone.

In short, solutions cannot be sought in technical and material revolutions outside man, but must be found in a reformation of humanity from within. Wisdom and morality are essential. We must realize first that relying too exclusively on mechanical devices like the automobile can cause grave regression, even atrophy, in our own muscular functioning power. The man who rides in a car all the time loses the ability to walk long distances vigorously. Even when we employ mechanical devices, we must realize that we human beings are in control and that whether the machine is a blessing or a curse depends on what is inside us. The most advanced high-performance automobile will not function properly unless the driver knows how to operate it (wisdom). That same fine automobile can become a coffin on wheels unless the driver has the good sense (morality) to operate it cautiously at safe speeds.

During his modern history, man has been deluded into believing that the key to happiness lies in reforming exteriors. The consequence of the misplaced emphasis on the exterior has been neglect of – even total oblivion to – the inner workings of human life, the need to suppress some mental actions and encourage others. Man's most pressing task today is the elevation and reformation of his inner spiritual life. This is what I call the human revolution.

Is conscious self-awareness sufficient to the accomplishment of this revolution? Obviously, being aware of the need for change is better than being unaware. Still, restraint from without and impulses from within create circumstances beyond the control of will or conscious awareness. As depth psychology points out, in the lower levels of our minds stretches out a vast world of subconsciousness from which occasional powerful impulses beyond the control of rational judgment emerge. The human revolution is impossible unless this subconscious world is altered. Religion is the most important single factor in revolutionizing a human life for the reasons I shall set forth in the following paragraphs.

My own conviction is that practical application of the teachings of Buddhism is needed to revolutionize our subconscious world, because these teachings are imbued with the power of the greater universal life behind all individualized, manifested life. Buddhist doctrines can alter human destiny. A clear, rational explanation of why this is true is difficult to make.

125

To try to clarify this important matter, I offer the following brief treatments of the Buddhist doctrines of the nine stages of consciousness and the ten realms of existence.

The theory of various kinds of consciousness was evolved in its fundamental form in the fourth and fifth centuries by the Indian scholar Vasubandhu, though it was later modified and the number of consciousnesses was increased to nine. The first five of the nine consciousnesses correspond to the organs of sense perception – eyes, ears, nose, tongue, and body (by which is meant percepters in the skin). In addition to these, there are four kinds of consciousness related to the mind. The first and most shallow of these could be called the ordinary consciousness, which organizes and coordinates information received through the sensory organs. The next, which is called the *mana* consciousness, deals with mental activities independent of external sensory information. Still deeper is the *alaya*, or storehouse, consciousness containing all memories from past existences that never reach the level of ordinary conscious awareness. The *alaya* consciousness contains all the karma-controlled elements determining spiritual and mental characteristics and fate in this life.

These first eight consciousnesses formulate individual characteristics and manifest themselves in the phenomenal world. The ninth consciousness, the *amala* consciousness, is the basic level, containing the true universal reality, or life in its universal form. It is called the undefiled consciousness because no effects from past acts stain it. Though pure in itself, the *amala* consciousness can, through external matters – especially relations with other living beings – produce defilement.

The *alaya* consciousness is comparable to the subconscious set forth by depth psychology. The *amala* consciousness, however, is deeper and more fundamental. For this reason, if an individual life is to be revolutionized it must be on this fundamental level, since human actions depend not merely on the functions of mind and body – the first eight consciousnesses – but also on the ultimate reality of the universal force of life, which is found only at the *alaya*-consciousness level. Buddhism teaches two approaches to such reform. One works changes in the characteristics and fate determined on the *alaya* level. The other is a basic development of the *amala* level itself.

Buddhist though analyzes human happiness and unhappiness in terms of ten realms of existence: Hell, Hunger, Animality, Anger, Tranquility, Rapture, Learning, Realization, Bodhisattvahood and Buddhahood. The state of intense misery, Hell, contrasts with the state of maximum liberation and happiness, Buddhahood.

As might be expected the ten states are intimately interwoven with the nine kinds of consciousness. Hell, Hunger, Animality, Anger, Tranquility and Rapture all involve happiness or unhappiness produced by external causes and are therefore dependent on the first six stages of consciousness. The states of Learning, Realization and Bodhisattvahood are those realms in which the self attempts to establish its independence; these states do not depend entirely on exterior factors and are therefore related to the seventh or *mana* consciousness. Freedom from external causes demands operation of this consciousness, but this alone is insufficient to total liberation, since many other constraining elements are found lower in the eighth, or *alaya*, consciousness. Overcoming the evil in the *alaya* consciousness requires the building up of good powers of fate through good deeds. Persistently doing good is called the discipline of the Bodhisattva. Since evil karmic elements have been stored up through the distant past, balancing them with good karmic factors through good works takes a very long time. This is why Buddhist doctrine holds that the discipline of the Bodhisattva takes tens of thousands of rebirths.

A Buddha is one who has understood the *amala* consciousness directly, and in his actual living is bound by neither external factors nor internal impulses. He is thus totally independent. As I have said, some schools of Buddhist thought advocate revolutionizing the inner beings by means of changes on the level of the *alaya* consciousness; such a revolution takes a very long time to achieve. Only the *Lotus Sutra* teaches an inner revolution affected by developing the *amala* consciousness; that is on the level of the fundamental universal life force.

Such development on the *amala*-consciousness level is comparable to working and fertilizing agricultural soil. The inner revolution only bears fruit, however, when it is directed outward in practical acts in actual society. Such acts can be compared to planting and raising crops in that prepared soil.

The person who has undergone a human revolution on the deepest level of his consciousness knows how to relate with nature and with other human beings. He knows that man must strive to remain in harmony with the forces and cycles of nature. Furthermore, he knows that respect for the existence and rights of others and for the dignity of humanity – Christian love and Buddhist compassion – must be the basis for all personal and social relations.

Second in importance to religion in bringing about the human revolution is education. Unfortunately, today, education leans too far in the direction of imparting bare knowledge and neglects instructions in basic attitudes toward all human life, including knowledge. This is a serious drawback, not only from the children's standpoint, but also from the standpoint of society and all civilization because knowledge in the hands of people who do not know how to use it can be either of no value or detrimental. All too often, universities and other institutions of learning teach their students information of direct use only to people who pursue the given specialist field after the termination of their academic careers. For people who engage in other lines of work, their hard-earned knowledge is forgotten after graduation. It is valueless to them, though it is true that the discipline entailed in the learning process itself may be useful. The amount of valuable learning retained by young people today is disproportionately small in comparison with the energy expended in acquiring it. Moreover, for the sake of imparting this knowledge, which may or may not be fruitful later on, we sacrifice truly important instruction in basic human values and attitudes. I feel that a reform is needed.

I do not reject the importance of knowledge. Man must know many things to live in society. He must study the past to be able to cope with the present. He must learn about the geography of the world in order to understand relations between his own region and others. As Tsunesaburo Makiguchi, first president of Soka Gakkai, pointed out, knowledge about a child's immediate geographical region, its climate, industry, social organization, and culture, leads to interest in other more distant regions. By learning where his own district fits into the whole country and by finding out where the food he eats comes from, the child widens his vision to include the entire world. He acquires knowledge of things and places that were unrelated to him before. This knowledge remains with him, it is never useless. Finding out about the political structure of his hometown, then about that of the nation and those of other nations, enables the child to understand his own position in the immense overall structure. Learning how important small living creatures are to man inspires curiosity to learn about the whole miraculous world, and cultivates respect and gratitude for the wonder of life. Step-by-step growth in knowledge of this kind exerts an inner revolutionizing effect.

To recapitulate, I think the human revolution consists of two major parts. First is the application of religious (in my case Buddhist) practices to the development and improvement of the innermost levels of human life and consciousness. Second are contacts and practical activities – related to religious practices and such things as education – in the world of actual society.

When the human revolution is achieved in the inner and outer beings of more and more people, human relations and relations betweeen man and nature will be more harmonious. This will provide a reliable basis for the solution of the grave problems – environmental pollution, war, exhaustion of natural resources and so on – currently facing mankind.

THE HUMAN REVOLUTION

Views of Aurelio Peccei

The end of a century has always struck people's imagination as a milestone in human life. Now, we are approaching such a great moment – one that coincides with the end of an extraordinary millennium that has seen humanity struggle from the dark ages to the threshold of a new era that promises to be even more extraordinary and fateful. We feel ourselves to be at *one of the great turning points of history*. While in the past the ebb and flow of human fortunes was represented by the slow rise or decline of one civilization or another, today changes are likely to sweep across the entire globe, bringing uprecedentedly good or bad fortune for one and all.

The future seems to be full of promise and, at the same time, fraught with more danger than ever before, for our generations have acquired the knowledge and power both to open up new wonderful horizons to human endeavor and to destroy themselves in an explosion of anger and violence. We are beginning to realize that what happens from now on depends on us to a degree never previously imagined, that we need new courage and foresight, and that we can no longer afford to make major conceptual or political mistakes. For the first time, we are face to face with our destiny – perhaps with that of the entire human species – which we must shape directly for better or worse by our behavior and deeds.

The emergence of this new sense of primary challenges and responsibilities is no doubt a good omen. It is not, however, enough: we must realize that the present trend of events must be corrected. Contrary to general belief, *humankind is on the wrong course.* In judging human affairs, we are still mesmerized by the perspective of a life of plenty bought cheaply, an illusion we inherited from the great feats and great expectations of the 1950s and 1960s, decades of unprecedented scientific discoveries, technological progress, and down-to-earth accomplishments that substantially improved the human lot. Advances were made in one field after another – in the conquest of space and the mastery of physical elements as well as in medicine and in education no less than in industry, transport, and communication. An air of euphoria about the future prevailed, and prospects seemed good for a higher standard of living and a more wholesome existence for rich and poor alike. Everybody was too engrossed in enjoying this foretaste of the benefits of novel techno-scientific prowess and economic performance to fuss about future costs or limits, let alone to consider a possible downturn in world conditions.

The need for a warning against the blind self-complacency of our material civilization was, and still is, part of the Club of Rome's *raison d'être.* The means initially chosen to voice concern and to send out a call for a greater sense of reality was the well known report on *The Limits to Growth,* published worldwide in 1972. Although controversial, this report had a notable impact on public opinion. Illusions, however, die hard. Though a growing number of people now realize that the multiple crises we are currently grappling with are a sign that the institutions, policies and strategies that earlier succeeded in keeping the human system reasonably well on the move are no longer able to accomplish this function adequately, the hope that an unexpected Eldorado may appear around the corner is still widespread. Most governments and political parties, and the world establishment generally, still nurture this false hope, while at the same time desperately searching for ways and means of

132

accomplishing the impossible task of turning the clock back and recreating the boom atmosphere that existed in the halcyon days of a couple of decades ago.

The new step that must be taken is to look at the situation as it actually is and to assess its likely medium-and long-term evolution as realistically as possible. I am sure that an objective analysis of this kind will enable us to realize that we are traveling in the wrong direction and that it would be a fatal mistake to persevere in what we are doing – sticking to the same lines, methods, and mechanisms that, in matter of two or three decades, have landed us in such trouble.

Such an analysis would certainly make a few basic matters eminently clear to our spirit. First of all it would reveal the utter folly of rushing blindly ahead, trusting that more and more technological and material progress can solve all human problems, or expecting the world's degrading situation to be redressed, as if magically, merely by instituting a new international economic order. Many people continue affirming this point of view in more or less good faith, although nobody has yet convincingly defined what this new order should be or how it can be brought about. It is equal folly to rely on the possibility of reaching our objectives simply by putting the arms race under control and quickly jumping into a new age of plentiful energy provided by nuclear fission or fusion, as others maintain, or by bringing education – what education? – to every village and hamlet of the world.

Even if some of these developments have a number of positive effects and are probably bound to occur anyway, the reductionist approach of relying essentially on a specific fact or measure to resolve the multiform human crisis, is less a solution than part of the problem itself. At heart the average citizen sense this and fears that, in addition to announced benefits, there are going to be hidden costs and drawbacks almost impossible to estimate in advance. The time has come to say bluntly that what

133

we are doing today actually boils down to little more than dealing with the symptoms and manifestations of our predicament, rather than with its causes, and that by 'staying the course' we will probably end up being drawn deeper and deeper into a vicious vortex of decline. The gravest consequence of our fallacious belief that our current dealings, policies and strategies will eventually bail us out of all crises, is that this belief diverts our attention from the real core problem. Something else is at the root of our crisis, something intangible and still undefined, yet fundamental, which is within ourselves and is so great as to generate the otherwise incomprehensible ills that possess us. This is *our state of inner disorder*.

This human weak point is the real Achilles' heel of the personality of modern men and women and, hence, of our triumphant civilization. It stems essentially from our reliance on the outmoded and erroneous concepts of ourselves, of our world, and of our place in it that drive us to think and act incoherently, in ways that are at odds with the real world – ironically enough, as we increasingly interfere with and transform everything on the planet ostensibly to make it all better serve our needs and wants. Our individual imbalances, uncertainties and tensions tend to spread to our communities and societies and then, multiplied, rebound on us as individuals.

The processes that are both cause and consequence of our internal disorder and growing mismatch with our environment have been stimulated by material revolutions that have brought about phenomenal changes, setting this day and age apart from all those of the past. Thanks to these changes, we have suddenly acquired unexpected knowledge and power. We must recognize, however, that, inebriated by it, we mistake all this as a sign that we are indeed the hub of life. Quite evidently, alas!, *our new knowledge and power are not accompanied by new vision and wisdom*; all too often we abuse them or use them extravagantly, creating in our total environment serious mutations that we are then unable to control or unprepared to live with. The more we expand our knowledge and power, the greater the danger we may be in.

Moreover, we are tempted to employ our novel strength almost exclusively in the pursuit of material welfare, forsaking any spiritual or ethical inspiration and ignoring the best of our moral, social and aesthetic talents. By focusing essentially on things that, though enjoyable, are outside us, we neglect our inner world and aggravate its faults and imbalances. All the traditional experience, accepted frames of reference, manners of judgment, and conventional wisdom accumulated over the centuries are being upset and put into question, while nothing of truly great value fills the vacuum. The pervasive crisis we are grappling with is the envenomed product of, and thrives on, our own innermost disorder.

We must become aware that the otherwise wonderful mechanical and electronic civilization we are tenaciously fostering has shaken our age-old cultural foundations, but has failed to show either capacity or propensity to reorder and reinforce them or to renew them in other, adequate manners. This awareness is now surfacing in many people. Sensing the perilous straits in which we have trapped ourselves, we begin to perceive how badly we need to find the strong support of new moral and intellectual energies, and the self-reliance required to reconstruct our inner world as it was − a safe haven of refuge and a trusted guide for action in time of trouble. Only a new humanism, both uncompromising in its ideal motivations and congruent with today's technical realities, can support us in this epochal reconquest of ourselves. Only such a humanism, where salvation starts within ourselves, can give us the force to reach higher thresholds, allowing us to scan alternative paths toward the future. This renaissance of the human spirit in a time of great distress is what I mean by the term *the human revolution*.

The concept of a revolutionary renewal from within is not

purely utopian: it responds to the primary need to survive, to avoid self-destruction. At the same time, it represents a kind of cultural evolution that our generations are in a condition to undergo. It belongs to the domain of the *Real-Utopias* on which we must rely to get out of our present predicament. The slowness and confusion with which this concept is emerging is not due merely to its complex configuration. One of the reasons is that, in spite of our awareness that we live amidst a great many man-provoked mutations, we remain reluctant to recognize that these mutations deepen the human plight today and tomorrow. We are even less ready to admit that the major changes required to offset their adverse effects must take place within ourselves. Whatever the causes and however paradoxical it may seem, this is proof that we must complete a full apprenticeship in forging ahead in this brave, or not so brave, new world.

The key lesson to learn is that we must live in close harmony with our part-natural part-artificial environments and are therefore not at liberty to modify them at will. This means that, whenever we meddle with natural systems or superpose our new ones on them – something we do practically every day – it is essential to exercise utmost restraint and a sense of responsibility, making thorough evaluations of all possible consequences for the outside world and of our own capacity to adapt to them. Besides promoting our own development, the human revolution should aim at enabling us to devise mechanisms of mutual adjustment among man, society and environment. *To live at peace with nature* has become a matter of urgency not only for the reasons already expounded, but also because maintaining peace with the outside world means that we must necessarily live in peace among ourselves as well. I shall return to the intrinsic nexus between these two indispensable states of peace, which are a premise for any sane reasoning and safe living from now on.

Other requirements must also be taken into account if we are to understand – and adapt to and finally control – the complex transformation processes that we are causing and that affect our lives for better or worse. Certainly, we must consider the whole of *the human system in its global span.* To analyze some nations only, such as those of Western Europe or Japan, as we are often tempted to do; to focus essentially on the two superpowers, as if their alternatives could subsume those of all humanity; or to suppose that any nation or group of nations can in some way be isolated from the rest of the world can only be misleading in a situation of growing global inter-dependences. Equally mistaken is our tendency to separate problems of the economy or of energy resources from those of military security or the world's power structure, as if any of these issues could be tackled independently from the others. Each problem, or family of problems, must be examined in a wider context containing many other problems. A unitary view of the entire human system is necessary if we want to understand and cope with the new category of macroproblems that already exist and that are going to be even more numerous and threatening in the future.

We must also adopt *the long view.* At a time when events are unfolding faster and faster, to focus on the immediate or short-term results and sweep most of the rest under the rug, as we are doing now, can only pile up even greater difficulties and crises for tomorrow. What is necessary is continuously to assess the repercussions and effects in a more distant future – at least as far into the future as we can see – of what we do or fail to do today and of our ways of doing it. These longer-term considerations must not be limited to only slightly corrected projections into the next few years or decades of current trends, because we are probably about to enter a period of marked discontinuities, most of which cannot be anticipated in the sense that we cannot foresee when, where, and how they will happen. We must remember, however, that some changes in trends are to be expected and that they can influence our future appreciably if not decisively. We must be prepared to take these contingencies into account as something likely to occur, and devise ways and means to include them in our thinking as far in advance as possible, while we try to assess the probable delayed effects of our action or non-action.

Then there is the question of *the fast-growing tangle of complexities* we are confronted with every day everywhere in the world. By relentlessly expanding and refining our artificial systems, we have complicated them enormously. A whole range of new diversifications, dimensions, speeds, interrelationships, overlappings, tensions, instabilities and uncertainties severely test the skill and ingenuity of even the top experts, and confront ordinary people with challenges quite beyond their comprehension. When the human crisis is considered in its alarming depth, one of the key parameters to be taken into account is the threshold of complexity the average citizen can sustain and adapt to, and whether and when we transgress this threshold.

During the course of evolution, human beings have always had to cope with limits and restraints – of a material, as well as of a religious, political, social and psychological nature. These limits and restraints have varied depending on conditions, or in response to distinct concepts of good and evil as defined by different cultures, and have formed the subject of reflection since time immemorial. The complexification of the modern world is so great that we risk being caught up in a maelstrom of confusion which will make us totally incapable of finding our bearings. To know something, we must know it in a simple form. A supreme effort is necessary to devise ways and means to reduce complexity to simple terms – as the philosophers and thinkers of the past have done.

Never before has it been necessary to construct an overview of all the world's regions and all human problems and possible solutions, and to take into account their basic interrelatedness and long-term projections. Therefore we must now do something original, something that has not been attempted so far. We should not, however, be taken aback either by the need to attend to many requirements and to consider many parameters, or by the fact that the process of analyzing the nature and dynamics of the global systems and arriving at a stage where we can exert reasonable control over them can be developed only gradually. Nor does it matter much if we take time getting into our stride, if our initial appraisals contain a number of gross approximations, or even if our findings or conclusions remain purely preliminary in character for quite a while. *The important thing is to start working in the right direction* for, if we succeed in doing this, favorable developments will pick up momentum by themselves.

We must not remain in our present state of ignorant confusion and apathy in the face of the factors and phenomena that are responsible for our current global crisis and whose interplay can further aggravate it. We need to be able to recognize, assess, and monitor these factors and phenomena. This can be achieved only by an in-depth, holistic, planetary, far-searching approach. Of course, thousands of other things must be done on all fronts practically everywhere, to alleviate local and regional problems and to meet other manifold exigencies. Nevertheless, the most indispensable act is discovering where we are at the global level and where the impulsion of present-day trends may lead us. Only a clear perception of the wrongness of our path and of the magnitude of the danger threatening everything we have and everything we are, can mobilize the willpower and energies we need to change course. Only this can stimulate the formation of the great synthetizing visions, motivations and ideas capable of inducing a multibillion, heterogeneous population to accept the risks and sacrifices required to launch humanity on a new course.

Once we start looking at our world and ourselves in this way, a profound cultural evolution will be set in motion. We will finally be compelled *to take a second, hard look at ourselves*, the protagonists – at our inner state of confusion, and at our poor performance. Reluctant so far to embark on a critical self-analysis, we have shown little interest in exploring why, in this golden epoch of scientific excellence and mass education, we have allowed the very existence of humanity itself to be put in jeopardy.

No matter how much we may like to quote the old tenet of wisdom 'Know thyself', the task we have to undertake now is far from easy. Obstacles other than our lack of propensity for self-introspection are numerous: misplaced national or ethnical pride; a certain tendency of the governing classes not to let the governed learn too much too quickly; sheer ignorance of the possibility of improving human quality, just as the quality of pigs and chickens can be improved; paucity of facilities for conducting such an enquiry – and many more. Ultimately the force of things will push us to making a thorough cross-examination of ourselves. And then, besides casting light on our deficiencies and mistakes, we will discover – to both our great shame and to our happy surprise – that *within ourselves lies a prodigious wealth of undeveloped and unused capacities* that have never been so much as explored. There exists in each individual a natural endowment of qualities and abilities that have been left dormant, but that can be brought out and employed to redress the deteriorating human condition.

As The Club of Rome has repeatedly affirmed in the last few years, this forgotten yet available human potential appears to be so rich and promising that its enlightened development can probably be – no, certainly is – sufficient to offset all our weaknesses and limitations. It can become the trump card up our sleeve that will help us turn the tables. The innate, vital resourcefulness and intelligence intrinsically inherent in every human being – from the most talented and fortunate to the most deprived and marginal – constitutes the unequalled patrimony of

140

our species, even though we are currently squandering and misusing it badly.

There is no doubt that our latent capacities of comprehension, imagination, compassion, solidarity and creativity; our basic learning ability; our neglected skills and our other still untapped positive qualities and attitudes, as well as the moral energies that are part of our very humanity, can be stimulated, groomed, developed, and mobilized to good ends. They are waiting to be called forth. This dormant human potential is scientifically recognized. It exists within our brains and can be retrieved from its present state of abandonment to become our greatest resource – a most extraordinary resource indeed, one that is both renewable and expandable. The primary goal of the human revolution, and by far the most important outcome to be expected from it, is *the full development of our innermost capacities.*

Viewed in a broad anthropological sense, therefore, human development is emerging as the most basic imperative of this new age. Its very concept of cultural promotion and advancement of the entire human personality is larger and more ambitious than that of development in its current application – even if the adjective *human* is often appended to the word *development*. The concept inspiring our development aims is usually associated with the concept of human needs and their satisfaction, and is therefore essentially utilitarian in character. Giving priority to need-oriented people-based development in the present-day world is understandable, because the conception derives from the struggle waged by the disadvantages in every nation and by the so-called international proletariat at large, in the hope of achieving a better life and a more equitable society.

Envisioning cultural development requires a more insightful and sedate analysis of the various aspects of the human predicament and is therefore much more difficult, although indispensable if we are finally to achieve the kind of development that is dearer to the poor.

141

I submit in substance that the development of the overall human potential is fundamental to progress of human society. It is both conducive, in the course of time, to what we now call economic or socio-economic development and uniquely able to provide our economic activities with the ecological foundations they must have in the future. In fact, only by developing ourselves culturally can we really understand how essential it is to reestablish our relations with the environment on healthy, sustainable bases – that, in the long term, will be economically rewarding as well. Pursuit of the brand of economic development we are currently seeking, on the other hand, all too often fails to consider the environment, thus transfering the cost of its rehabilitation to a future account.

Other reasons, of an ethical, political, and psychological nature, press us to strive for all-embracing human development. The most important perhaps is that such development alone can furnish people at large with the stimulating challenge and satisfaction of taking an active part in the conduct of human affairs. This in turn will permit people to express their personalities in their societal environments while contributing to the common good.

Well balanced cultural development represents a pre-condition for all other types of development, since, quite evidently, no development, economic, social, technological or institutional, is possible or can be meaningful if the people themselves remain underdeveloped. If we are at long last to stop the runaway course of the material revolutions and start controlling, blending and guiding them toward chosen ends, a full-fledged, human revolution based on the uplifting of our qualities is now indispensable.

It goes without saying that the things required to set a teeming humanity on the path of self-development are *momentously difficult*; nevertheless, they are also *fundamentally simple*. This crucial point must be understood. For all its novelty and formidability, the enterprise of raising human capacities above their present level still belongs to the realm of things within our reach. The reason for this apparent contradiction is that all the factors required to spark off and then carry on our own development exist, embedded in our mental endowment, but have so far been ignored or neglected. The whole enterprise thus depends essentially on our ingenuity and artistry, which have given proof of their worth on other occasions when we were in dire need of them. We must now have recourse to them once again, to conjure up and combine our other capacities, so as to engender the process of acculturation that can make our generations a useful link between those that have preceded us and those that will follow.

I have described elsewhere[1] the feasibility of this enterprise and have sketched in a few guidelines that could get it moving and will not dwell on the matter here.

Nonetheless, I wish to issue a warning to the effect that such a lofty goal can never be attained and that our race toward a dark future will continue if we become dispirited and lose drive. How can we change our condition if we are convinced that modern man cannot transcend his present inner limits, if we are persuaded that he is therefore incapable of liberating himself from the prides and prejudices that have brought him to the edge of an abyss? What we think about our own capacities and possibilities is all-important. To presume that the lines of modern man's drama are already drawn for the future and that it is in vain to attempt to modify them is tantamount to accepting defeat before trying to do anything about it.

We must combat such retreatism and become realistically aware that the human resource, so abundant and ubiquitous, can be developed and deployed to make our venture fully

[1] *The Human Quality*, Pergamon Press, Oxford, 1977. (In German, *Die Qualität des Menschen*, DVA, Stuttgart, 1977).

worthy and increasingly more rewarding. 'The conviction which must enter the minds and hearts of ordinary citizens as well as opinion leaders and authorities everywhere is that at this decisive juncture the primary and most urgent obligation and interest of world society – and, at lower levels, of the nation, community, company and family – is to improve by all possible ways and means the quality of all its members, modernizing, upgrading and cultivating their individual and collective preparedness and fitness for the difficult times ahead. Whatever talent and treasure is necessary should be invested with absolute priority in this task of human promotion and development.'[1] The world needs confidence in itself and in its capacity to shape a better future. The development of our own capacities is the natural springboard to building this self-confidence and hence to realizing that the jump in human quality and qualities that is required to get us out of the pit is quite possible. *It is possible – if we wish it.* Acquiring this conviction will be both a premise of and a prelude to taking a constructive line of action from now on.

One of the major benefs to accrue from the human revolution is that, for the first time, the way to veritable peace will be clear. Thanks to a cultural renewal, we will be able to cast the crude, critical light of common sense and morality on our gigantic military and security establishment, revealing the hallucinations whereby it has distorted reality and pampered and exploited our less noble passions, thwarting all possibility of instituting peace in human society. This new light will permit us to discover the genuine and valid, if rigorous, conditions that six or seven billion inhabitants of Earth must set themselves, to live and let live in the decades to come.

The gestation of a truly peaceful society in a warless world can be imagined only as a long and laborious process. The

[1] *Ibid*, respectively pages 148-49 and 170-71.

conflict-prone society of today still deludes itself that security can be bought by investing in modern armoury and that the alchemy of intricate disarmament deals can conjure up peace, even as society is prepared to wage war if problems appear insoluble or if the booty is too tempting. To become truly peaceful, such a society must undergo a profound cultural maturation. The resulting new society will probably have to be begotten by the old one by an internal evolution similar to the metamorphosis of a caterpillar re-forming itself into a chrysalis, protected by a cocoon from which it then emerges as a butterfly – a process in which each stage already contains some traits of the following one. The main catalyst for this evolution will probably be peoples' growing perception of the abyss from which they are slowly emerging. Looking back, we will all wonder how the world could have been crazy enough to trust its lot to instruments of ultimate destruction, calling on the best talents to perfect them and make them foolproof for the mass-destruction of humanity. It will be the dawn of a new era when mighty modern mankind wakes up to see that, because of overwhelmingly size, its power can no longer be used in war. Willy-nilly, merely to survive, mankind will then be compelled to be peace-loving and peaceful. There will be no way of camouflaging war as something useful or patriotic: world peace will have asserted itself as an imperative beyond the shadow of doubt.

In the context of this broad reappraisal of our condition and of what it will take to make it safer and qualitatively better, we will probably be encouraged by the appearance, in the midst of so many difficulties, of new positive phenomena endowed with great momentum. I refer particularly to the spread, convergence, and alliance of the three major grass-roots movements aimed at stamping out material violence from human society. These movements in defense of nature, of peace and of human rights have sprung up all over the world as the result of action by groups of committed citizens and can be seen in many countries, tips of icebergs of huge proportions. A fourth movement is simmering or exploding almost everywhere, even if it has not yet been able to manifest itself in equivocal terms. Although numerically much stronger, this movement responds to an extremely variegated assortment of motivations and is even rent by contrasting interests that have so far prevented the efficient deployment of its full force. It is the worldwide movement for social and economic justice that I have already mentioned, and that will probably acquire greater cohesion and effectiveness only when the human revolution is in full stride and can underpin it with wider-based intellectual and more coherent political supports.

It is beyond the scope of this volume to discuss the prospects and possible impact of these movements in detail. Let me just say that their major significance and potential strength lies in the fact that they are not the brain-children of the ruling classes. They surge up from the roots of society and have a strong popular character, which shows that in spite of – if not thanks to – the disarray and fears besetting the world at the end of this century, the societal body of a great many communities still has an unsuspected amount of vitality. These movements are producing what in a sick organism would be called antibodies, namely, elements that fight the ills assailing it.

In conclusion, although these popular initiatives themselves are scattered and uncoordinated, they offer the most encouraging proof that there is a widespread awakening of public conscience and a growing demand for a cleaner, more just and more livable society – sound seeds for the human revolution.

There is, nonetheless, at the core of this revolution, a quintessential question that of late, in the rush of doing things,

[1] See also *Der Weg ins 21. Jahrhundert,* Molden, München, 1983.

we have forgotten to ask but that we can no longer dodge. This question synthetizes the uncertainty we have about our own finality as human beings, an uncertainty that will grow if we succeed in broadening our horizons. Where can and should we go, as humankind and as a species, if vision and wisdom guide our extraordinarily accumulating knowledge and expanding power? With such stupendous capacities, what should we do with ourselves and accomplish in our world? The very newness of our condition and its implications for everything else leave us aghast. I for one am a little frightened. Yet, I realize that it would be vain to try to ignore all this because, wherever we are, the key question will remain with us and will press for answers. I understand that it is impossible to put off facing this issue until we know more, because we must take responsibility at the highest conceivable level for both ourselves and our world if the human future is to be guaranteed.

For quite a few years now, The Club of Rome has been concerned with these problems. As they have worsened, we have felt that we must try to do something to confront world crises. We decided to initiate an examination of the alternatives humankind can realistically envisage during the next phase of its evolution. For this purpose, we have been encouraging an enquiry aimed at tentively detecting the *more desirable alternative futures* that can be attained in the light of the present-day situation.

This project, called *Forum Humanum,* puts the accent on the young. It is they who have both a greater stake in the future and fresher minds and hearts to devise ways to improve the world. In various countries, groups of young men and women committed to this enterprise are already analyzing the realities that count most in the world: those of global relevance and long-term impact, those requiring transcultural and transdisciplinary approaches, and those calling for both thorough reflection and enlightened action. Their objective is eventually to arrive at a

147

model of the coherent structures of a few of the more positive alternative apparently possible scenarios, and at an outline of what our generation must do to realize one or other of them. They hope to be able to present some conclusions on the occasion of International Youth Year, 1985. Even if this objective cannot be fully attained, it is essential for them to show that current options are so numerous that, if prepared to take the required steps, humanity can set itself on a much more favorable course.

By themselves, the work and pleas of these young people will certainly not modify the world situation much, if at all. Hopefully, they can show that another future can indeed be shaped for all the world's inhabitants and thereby inspire others to take the flag they have raised and go further in the search for safer paths and worthier goals for the years and decades to come. In this way more seeds will be sown for the human revolution and will eventually come to full blossom.

The age-old question of what destiny holds for humanity remains beyond our comprehension and probably can never be answered. Nevertheless, the human revolution as I have described it is indispensable, for it alone can give us a much higher point of observation from which to look ahead and enlighten us about what may lie in store for humankind. It alone can help us understand that, having grown so much in number, sapience and might, filling all the spaces of our terrestrial abode and mastering everything in it, we must now shoulder, for the first time, long-term global responsibilities and strive to leave a more livable planet and a more governable society to coming generations. It alone can make us realize that, to accomplish all this, we must improve within and acquire an inner spiritual and cultural harmony. Thanks to this human revolution, the end of our century and of the present millennium can become the door to one of the best periods in human history.

1. Inside, Not Outside

IKEDA: The grave world situation has led you to call for a humanistic revolution. Basing my thought on Buddhist teachings, I call for a human revolution, which is a way to change the inner nature of each human being and thereby bring about improvements in society as a whole. Through developed scientific technology, human beings in industrialized nations have created a way of life so materially rich that kings and aristocracies of the ancient and medieval past might well envy them. Yet man has used the same sophisticated technology to produce weapons of unprecedented destructiveness – weapons that have already taken a heavy toll in injuries and lost lives. Pursuit of material well-being has robbed man of value and of the warmth of human exchanges. In this way, it has bred violent crime and despair. Instead of employing the immense power of science for the happiness and enhanced dignity of all mankind, a limited number of people and groups use it selfishly to satisfy their own desires or in competition with each other to ascertain their superiority over each other.

Buddhism refers to the mental states behind such misuse of scientific power as the Three Poisons (greed, wrath, and folly) which Buddhist practice is designed to enable people to control. In some cases, reason alone can alter the workings of the human mind, but many deep mental attitudes and operations lie beyond the control of reason. To deal with them, Buddhism insists that the practices of faith are necessary. My idea of the human revolution is based on such practices. What kind of practical action do you propose for the attainment of what you call the humanistic revolutions? Do you have a religion or religious faith in mind?

PECCEI: When I affirm the need and call for a human revolution, I do not refer to any religious faith; as you know, I have in mind a profound cultural evolution inspired by a new humanism capable of illuminating and inspiring our generations in this materialistic-oriented technological age. My discussion with you has nevertheless shown that, although you are guided by your Buddhist faith and although we start from different viewpoints and use different forms of expression, we are both talking about the same kind of change in the human heart.

What I maintain is that we all need to correct our vision of ourselves, our world and our place in it, and hence of our way of thinking and mode of behaving. For quite a while now we have been acting under the overwhelming influence of the material revolutions, which have permitted us to impose our yoke and our whims globally at will, thereby strengthening our inherent belief in the absolute primacy of our species. We have thus become dangerously self-centered and self-righteous and altogether so psychologically imbalanced that we have lost every possibility of making any serious critical analysis of our true condition and behavior. I have explained that, in my view, this is the main cause of our predicament.

To liberate ourselves from this fatal handicap so that we can make an overall evaluation of what has happened, or may be happening, to us and concurrently to undertake an objective self-analysis to discover our faults and errors we must first *overcome our infatuation with the material revolutions.*

In this connection, it is useful to note that ordinary people see the techno-scientific and industrial developments as a exogenous

150

phenomenon in which they have been involved, but which was engineered outside them by small groups of researchers, inventors or entrepreneurs, and gladly welcome it for the benefits offered, while accepting concomitant inconveniences as the price of the benefits. People in general were never able to embrace in thier vision the multiple processes that these events were bringing in their wake or to understand their final purport; they had no alternative, therefore, but to expect that all this 'progress' and those responsible for it would continue to carry on, in the right direction.

Only recently has the appearance of the patent excesses and negative side-effects of our modern civilization begun to shake the blind trust heretofore placed in progress and in those who were supposed to know how to guide it. And, as is the case when one starts to examine a complex situation in greater depth, we discovered all sorts of things – some good, some not so good and some really bad – that had previously gone unnoticed. It thus became evident that, in some instances, the price the world is paying for the benefits derived from the material revolutions is far too high. Furthermore, it emerged that quite a few of these instances are by no means marginal, but have a very major bearing on humankind's future – happiness, quality of life and even perhaps its very survival.

We are all beginning to perceive how unbalanced our condition has become. We have the feeling that we ourselves lack internal balance and that to get materially richer we have in a way become spiritually, morally and philosophically poorer. We are starting to see how perilously lopsided our societies are, for even our material benefits, welfare and progress are most unevenly distributed among and within nations, and the fate of billions of people lies in the hands of very small privileged elites who call the tune that, willy-nilly, everybody else must dance to. It is now dawning on us that we are dangerously on the wrong foot with nature too, and can no longer rely on it to support our sprawling enterprise and absorb all we are doing to make our position ever more formidable. As we perceive all this, realizing that never before has the world been torn by so many marked disequilibria, we hope and pray that something can be done to re-establish a modicum of dependable balance within our inner selves, within our societies and within our environment.

IKEDA: It is certain that disequilibria within mankind have brought about the imbalance you mention in society and the environment. Many people are trying to restore balance, but in general their efforts are directed to such stopgap measures as discovering a replacement for some natural resource that threatens to be exhausted or to correcting flaws in social systems by evolving new systems. It is safe to say that practically no one adds consideration of the need for a fundamental revolution in the way we human beings think and live. Retarding any changed or breakthrough in the prevailing situation, the mass of mankind is hurtling at a steadily increasing speed along a path to disaster. Scientific technology advances at a dizzying rate year by year, and each new achievement is greeted as something worthy of uncritical congratulation and unconditional joy.

We are like a fearless child infatuated with the delight of speed, stepping harder and harder on the accelerator of an automobile. A knowledgeable adult would take into account the possibility of curves or cliffs ahead and would regulate his speed to the situation and to his own ability to turn the wheel in times of unexpected crisis. The child plunges ahead, unaware that his situation is one demanding immediate change. In the case of mankind, that change is the human revolution.

PECCEI: It is none too early to start in earnest to redress the situation, not least because new waves of so-called progress are already looming on the horizon thanks to the rapid deployment and diffusion of microlectronic technologies, biotechnologies

152

and a host of other advanced technologies in such fields as space, the sea bottom and materials. These technologies are the end results of a science that is expanding fantastically and are essentially application-oriented, leading to developments (such as in information, robotization, and genetic engineering) that will have an almost immediate impact on society and all things human – social values, economic structures, political and cultural systems, individual attitudes, and so on. The damage they can cause if badly used may be irreversible (as in the case of nuclear warfare, elimination of species, and attack of nitrogen oxides on the ozone layers). Therefore, if caught in their onslaught in its present state of unprepardeness, humankind may well be swept into total chaos.

Extricating ourselves from the dangerous stricture in which we have become enmeshed by imprudently racing ahead in this way, without knowing where we were actually going or whether we will be able to control our speed and direction, is the great problem of our time. It is a problem that requires, first and foremost, a supreme cultural effort to understand why we have found ourselves in such a predicament, and then to form a clear vision of the remedial philosophy of life and practical action best suited to permit us to regain safe ground for our march into the future. I maintain that this indispensable evolution can be triggered and accomplished only be relying on *a strong, innovative humanism*. This alone can provide the countervailing inspiration and force capable of challenging, taming and then guiding to good ends our triumphant but rudderless 'progress'. Such a humanism is the essence of the human revolution that can give meaning and finality to the material revolutions we have adventurously launched, propelling humankind into a bewitching but hazardous new phase of its venture.

153

2. Ideals and Objectives

IKEDA: My concept of the human revolution originated with Josei Toda, second president of the Soka Gakkai.[1] During World War 11, Toda was imprisoned by the militarist authorities for his religious convictions. He insisted on living in accordance with the precepts of Nichiren Shoshu, the orthodox school of Buddhism founded in the thirteenth century by the great religious leader Nichiren Daishonin. In the 1930s and the first half of the 1940s the Japanese government demanded total allegiance to state Shinto, which they adopted as the mainstay of their version of the imperial system. They rigorously punished people who held other beliefs. Prison was Toda's punishment.

During his incarceration, he had a profound religious experience which inspired him to devote the remainder of his life to propagating faith in the teachings of Nichiren Shoshu. He called this awakening his human revolution and later explained that a revolution within the human being can be either good or evil. For instance, he interpreted the change of a pure young man into a creature hounded by the desire for revenge related in Dumas's *Le Comte de Monte Cristo* as a kind of human revolution for the worse. Toda's own revolution, on the other hand, had been for the better since it led him to devote himself to Nichiren Shoshu, and through it to helping fellow human beings find true happiness. As he said, to be good and meaningful, a human revolution must be based on clear convictions. Buddhism teaches that we must strive for an independence that, not governed by impulses of selfish desire or instinct, is in harmony with all things and, in a spirit of love and compassion, works for the happiness of all beings. The kind of self-alteration and self-improvement inherent in such independence is a matter of

[1] *The Human Revolution,* vol. 1 (1972), vol.2 (1974), vol. 3 (1976), vo. 4 (1982), Weatherhill, New York-Tokyo.

first concern to all Buddhist believers. The attainment of the goal is the achievement of Buddhahood. The process of development and discipline along the Buddha Way leading to it is the human revolution; that is, the human revolution is a process of advance from a selfish way of living to a way of living oriented toward the prosperity of society as a whole and of all living creatures.

The revolution cannot take place without some clear convictions. What are the convictions that you feel must underlie your humanistic revolution?

PECCEI: Certainly one of the objectives of the revolution I advocate is to prepare people better to cope with the complexities of the contemporary world and its artificiality, the new interrelations of everything with everything else, and the unaccustomed challenges it poses them individually and collectively – be they rich or poor or in the purlieus of the East, the West, the North or the South.

In past epochs, our predecessors undoubtedly had easier tasks to solve and could afford to make more mistakes; but they apparently succeeded in doing remarkably well on the whole, in spite of their more restricted knowledge and scantier means. Even looking a long way back, we are amazed at how primitive people who knew very little about their hunting grounds or valley, and nothing about what lay beyond, managed to gather enough experience and develop enough wisdom to subsist, making good use of the sun, rain, winds and seasons and getting along with all the plants and animals in the vicinity, including potentially harmful ones like snakes. We have transformed the Earth to suit our needs, eliminating or domesticating most of the other living creatures and crowding every nook and cranny of the globe with machines and devices that can be far more dangerous than snakes; but we have yet to learn how to live harmoniously with this changed world – or even with one another. Nonetheless, billions and billions of us must live in it

155

and must find ways to do so peacefully. Only the human revolution can teach us how.

However, *merely living is not enough.* Man must aim at being more than just a welfare-oriented biological organism, who fares well economically by exploiting nature's bounties, thanks to his ingenuity in devising all kinds of sophisticated and powerful artifacts. By birth, he is more than just a consumer and a producer. He is a spiritual, dreaming creature, who loves myths and seeks to communicate with his God; he is a playful being, a poet, an inventor, and an artist with immense curiosity and multiple skills, who deserves to be much more than the mere companion and master of his tools.

Only the human revolution can unearth our inner potential and make us feel fully what we really are and behave accordingly; only it can show us how to utilize our computers and satellites, our engines and instruments, our nuclear reactors and electronic gadgetry to commune better with our fellow humans and our entire universe. It is this revolution alone that can make us see how important it is to survive in order to have a life worth living, both for its own sake and as a means to prepare responsibly and compassionately a way of life for the generations who will follow us.

3. The Varying Revolution

IKEDA: For man to survive on this planet he must fundamentally alter both his way of life and his view of nature. He must stop regarding nature as something he can destroy and conquer for his own benefit, and must realize that the sole way for him to prosper on Earth is to realize that he shares life itself with all other creatures. Modern science and industry rest on the assumption that nature is there for man to use as he will. The very scientific approach dissects factors intricately intertwined in the world of nature, and applies mathematics to them for the sake of evolving generally applicable rules. To the extent that dissection destroys the whole organism, the scientific method can be regarded as nature-murder. Technology applies the knowledge science partially provides to extract energy for the destruction of some things and the creation of things that are totally novel. Pollution of the environment and spiritual desolation of humanity are two of the outcomes of the destructive conquest of nature.

Man tends to forget that, physically, he is constituted on the basis of the harmony and order of the matter of the natural world and that his mental functions reflect the harmony and rhythms of that same order. When he destroys or remakes nature he is exerting a destructive effect on his own mind and body. Furthermore, it is important to remember that the destruction taking place now will exert detrimental effects not only on the adult generations of today, but also on posterity. (Radiation effects from the atomic bombings of Hiroshima and Nagasaki are extending to second and third generations. The massive use of deforestation chemicals in Vietnam has resulted in malformed infants.) The mental effects of destruction of this kind are harder to assess than the visible physical ones, but probably none the less severe. How do you think we can alter our attitudes toward nature?

PECCEI: One of my foremost preoccupations is that we develop culturally, so as to realize that we must give pre-eminence to our relationships with nature and that this requires the adoption of a new ethic embracing life in all its manifestations. This, however, is an issue that cannot be isolated from the other major ones. Only through a holistic approach shall we be able to understand how fundamentally the human condition has changed and how imperative it is for us to be at peace with nature. To satisfy these requirements, we must evolve

157

from our present self-conceited egocentrism and develop the entire spectrum of our latent potential harmoniously.

Ours is a time of crisis, in which we must see man clear and whole, and in which our capacity to comprehend and create is direly needed if we are to prevent our crisis from becoming irreversible. Any major mistake in assessing our situation can be fatal. On the other hand, as I have often reiterated, the gamut of still dormant capacities that are available in each individual and that can be put to use is so great that we can make of it the greatest human resource. It is by grooming and developing these capacities in a way consistent with our new condition in this changed world – and only in this way – that we can restore a modicum of order and harmony into our affairs, including our relations with nature, and thus move safely ahead.

What is required is a cultural evolution that will be revo-lutionary indeed, for the new generous and forward-looking humanism which will have to prompt and sustain it must be strong enough to re-endow us with faith in this human being, placing him at the center of all progress. This profound in-novation should lead us from the present unwholesome stage – during which we are becoming even more knowledgeable and powerful while becoming morally and existentially weaker – to a new stage of self-fulfillment and accomplishment. With the renaissance of the human spirit, man will be stimulated to direct all his knowledge and power to the improvement of his condition, thereby becoming the true, discerning protagonist of his total venture.

To conclude, permit me some mildly provocative questions that underline the incongruity of our current stance. A great amount of research nowadays is aimed at the complex task of developing *artificial intelligence* and fifth generation, heuristic, 'intelligent' computers. With due respect for the worthy scien-tists engaged in this pioneering enterprise, I should like to know why we must pursue such a circuitous path to develop other kinds of intelligence, when natural and probably more important and rewarding intelligence is to be found around every corner in many men, women, and children. Why not concentrate on promoting *human intelligence*, which is waiting to be improved and refined? Are our expectations of the human being so low? Should not our political, scientific and religious leaders ask themselves whether it might not be better to have a world populated by somewhat more intelligent people instead of by highly intelligent supercomputers? And, if any of them are uncertain over their answers, whether it might not be better anyway to support research, at least equally, in both directions?

4. Guide, Not Master

IKEDA: Yes, I entirely agree with you, but I should like to pursue this line of thought further. Knowledge and technological power have apparently made man arbiter of the fates of all other creatures on Earth. Unfortunately, greed blinds man to his responsibility in this role; human beings continue to exploit other creatures, in some cases, right out of existence. Man may even bring about his own demise as a result of taking this course. To become a responsible arbiter on Earth, man must alter his view of life in all its manifestations. How do you think the humanistic revolution can help him do this?

PECCEI: In the wake of Western civilization, peoples everywhere have come to regard man as the master and arbiter of all things on Earth. Now that he is making his presence felt beyond the limits of our planet, it is inferred that he can rightfully stake his claims to outer space as well. In other words, what is good for man is held to be good for all things within his reach; and whatever he does wherever he can with a view to improving his lot is implicitly well done.

159

Both you and I judge this way of thinking to be preposterous and are convinced that its rationale is profoundly wrong and counterproductive, for *man is not the master, but only a part, of nature* – even if we contend that he is the most important part. The cold fact is that man is inextricably bound up with nature's innumerable elements and that, when he depletes them or impairs their cycles and systems, the damage boomerangs on him. What ultimately harms him most in this unimaginably complex symbiotic relationship is his cavalier behavior over non-human life. I will not repeat here what I have already said in this respect. Let me just add that, if our species has had the good luck to start its career with superior mental endowment, this does not give it the right to consider itself the supreme arbiter of whether less gifted creatures should live and, if so, how. Indeed, since we happen to have developed a brighter intelligence in certain respects, we ought to feel that this puts us under an obligation to be the protector and trustee of lesser forms of life.

Biologist and humanist Julian Huxley put this very well when he said that the role of man 'whether he wants it or not, is to be a leader of the evolutionary process on Earth and his job is to guide and direct it in the general direction of improvement'. This recognition of our responsibilities is part of the ethic of life I have already invoked. Apart from its fundamental ethical value, such an attitude must be recognized for its existential importance. As a matter of fact, the improvement of the texture of life on Earth no doubt also responds to the long-term interest of our own species – so much so that continuing to refuse to abide by this tenet puts the superiority of our humanity and our vaunted brainpower in question. Here then is a vast field to be taken up by the educational establishment of the whole world.

5. Their Way

IKEDA: The North-South problem – that is, discrepancies between the industrialized and the developing nations of the world – is certain to generate greater international tensions as time passes. It is undeniably the case that the conflicts that have taken place since World War 11 were caused by political instability resulting from poverty in developing nations and by interference in the affairs of these nations by industrialized segments of the world – the United States, the Soviet Union, China and Western Europe, for example. Lamentably, this interference further retarded independent growth in the developing zones.

The oil-producing developing nations *are* affluent, but the small amounts of their oil wealth used for the good of ordinary citizens is cause for the gravest concern. Instead of going to construction projects for the benefit of the people as a whole, oil money is too often spent either on private jetplanes and such frivolities as gold wash basins for privileged classes, or on military armament to be used in the fighting that breaks out from time to time among oil-producing countries. The policies of these same countries aggravated the already sad plight of developing nations who purchase essential petroleum at steadily increasing prices. This necessity and expense hamper their efforts to stabilize and adjust their political and economic systems, deepen already existing unrest, cripple attempts to cope with domestic famine and disease, and thus make the lot of the ordinary citizens totally miserable.

In my opinion, unless the people in charge of managing a national social structure are enlightened, responsible and intelligent, maintenance of order and good functioning cannot be achieved. The training of people qualified to carry out this difficult task requires a sound system of education.

Japan had a high level of education even before the

nineteenth-century introduction of Westernization and modernization. This is one of the reasons why she was able to close the
gap between herself and the industrialized West. Tremendous
enthusiasm for mastering new learning and technology inspired
the Japanese to expand and improve their educational system
further. How do you feel developing countries can obtain and
disseminate the education they require for the sake of true
independence and wholesome growth?

PECCEI: It is inappropriate in this heterogeneous world of ours
to use the word *they* to identify people different from us for, by
so doing, we lump together populations and cultures quite
different from one another and seem to attribute to our own
actions the quality of a model that others should follow because
of its special merits.

Even though I feel this way, I think that the developed nations
must stand ready to assist the developing ones with their own
experience and methods, whenever useful. Quite evidently, this
applies to a wide range of fields, but education is a very special
matter, for it is central to the cultural identity of all peoples.
While educational cross-fertilization may be very good, there is
no point in insisting that every country should develop its
educational system more or less according to the same pattern.
No matter how advanced the educational systems of the United
States or Japan may be, and no matter how much my country
can learn from them, Italy would err greatly if she were merely
to try emulate them. Similarly, though Italian schools of law, art
and the humanities have provided an example to quite a number
of nations, this does not mean that these other nations must
copy Italian ways.

Questions of education are, however, much more delicate in
the case of the Third World. Educational aid to developing
countries has frequently been no more than another form of
cultural imperialism; and, as a result, distortions in the donor

162

nations' educational systems have been transmitted to the recipients' systems. Moreover, the basic education of a highly industrialized country in a temperate climate is, to say the least, ill-suited for transplanting to a predominantly agricultural region in the tropics. Thus, scientific education cannot be transplanted from, say, Scandinavia to the Nile Valley, since the socio-economic priorities and educational needs of the two areas are totally different.

IKEDA: You are quite right. Education must train the generations who will bear the responsibilities of the future in their own regions. Since most regions differ from each other in social, cultural, historical and environmental backgrounds, each requires people educated to meet local needs. An educational system that is successful in one region may be disastrously unsuccessful in another. The intelligent people of each region must take their own conditions and future needs into consideration in evolving a suitable system of education – possibly by means of a series of trial-and-error attempts. No matter how difficult the task, since people constitute and maintain nations, the development of the right kind of educational system is of quintessential significance. Indeed, even if temporary shortages in connection with physical well-being must be tolerated, no effort should be spared in the name of education. Without a well-educated population, those nations that must rely on assistance from outside for food, clothing, and housing will have to go on relying on external help in the future.

PECCEI: To develop educationally or otherwise, the Third World should do its best to prefigure situations in which dependence on the North is lessened and in which not even North-South interdependence is the cornerstone. The nations of the South need something else: the enhancement of their own self-reliance. If they want to succeed, they must first take a hard

163

look at one of their own major shortcomings: their *fragmented political organization.*

At present the world is politically structured as 150-odd sovereign nations, the vast majority of which have the unsurmountable disadvantage of being small, weak and so riddled with problems as to be, individually, unable to discharge all the tasks incumbent on a modern state. Even if they want foreign aid, they are at a loss as to how to use it properly or apply it to complement their own indigenous capacity. Nevertheless, by and large the 120 or so less developed countries try to go it alone and are, consequently, no match for the political, economic and technological giants of the North – the United States, the European Community, the Soviet bloc and Japan. They are condemned to lag behind all along the line unless and until they find practical ways either of intimately coordinating their long-term policies or, better still, of coming together in regional unions, communities or federations. We can see this at a glance by taking the gross national product as a yardstick of a nation's capacity to perform. Even mighty India has a GNP about half the size of that of Italy, which is just a province of Europe. For almost all developing countries, the sheer fact of their inadequate economic dimensions and low per-capita income prevents them from doing so much as creating a modern physical and social infrastructure and upgrading their educational system, let alone holding their own ground in the give-and-take of international life.

If self-reliance is the path the Third World wants to follow, it must be a *collective self-reliance* of larger, relatively homogeneous geo-political or geo-economic areas ready to adopt common policies and institutions, an internal division of labor and a unified stand vis-à-vis the external world. If, on the other hand, the Third World believes it can develop satisfactorily and participate to a meaningful extent in world development – whatever connotation we may wish to attribute

164

to the concept of development – while remaining as atomized as it is at present, it will continue to reap frustration and setbacks.

As I have already mentioned, in 1980, a project was launched at the United Nations by The Club of Rome and two other future-oriented research organizations to help developing nations explore and exploit all available avenues for increasing their collective self-reliance by the establishment of organic groupings of nations bound together by structured, long-term arrangements. The enquiry is proceeding at a very interesting pace. Quite a number of skeptical political scientists or practitioners, who were reasoning only in terms of the classical framework of national states, are now convinced of the advantages for developing countries of uniting with a view to creating larger, more viable units.

As our discourse started with, and is still focused mainly on, education, I submit that education itself would benefit from these developments. It could in turn be instrumental in paving the way for the gradual restructuring of world polity into, say, a dozen or a score of continental, regional, or subregional communities with a capacity both for self-governance and self-reliance and for mutual cooperation in an interdependent world. Though directed principally at the young, education is a collective enterprise aimed at everyone. When all citizens have acquired and are ready to sustain ideas of self-reliance and cooperation, the whole world family will have taken a fundamental step forward toward a more wholesome future.

IKEDA: Internal conflict and warring among neighboring states is undeniably a major obstruction to the independence and fulfillment of developing nations. For this reason, it is, as you point out, vitally important to restructure world polity into continental, regional or subregional communities. Developing nations must put an end to foolish bloodshed among their own tribes and cooperate with their neighboring developing nations for economic and – most important of all – educational growth. Such medieval and renaissance European universities as Bologna, Paris, Oxford, Cambridge and Heidelberg are excellent models of institutions educating far more than the young of their own land alone. I believe the developing nations of the present can learn much by examining the ways these schools operated. Each was characterized by the fields in which it excelled; students gathered in them not according to nationality, but according to field of interest. Experiences and exchanges in universities of this kind prepare young people to be less parochial in viewpoint and provide a basis for negotiating on a wider plane of involvement when the undergraduates grow up to become the leaders of their individual nations. Leaders trained to think on an international plane in school are more likely to find amicable ways of settling disputes in politics.

6. Obligations First, Rights Later

IKEDA: Modern education sometimes concentrates on human rights and plays down human obligations, especially in connection with relations between man and nature. A man who owns a forest has the right to cut it down, but also has the obligation to restore the damage his act does to the environment. People today are too often taught to see rights and allowed to overlook obligations. I believe that the younger generations must learn that the two are indivisibly bound together on the individual and social scales. Do you agree with me?

PECCEI: I fully agree with you. There can be no rights without obligations, and obligations without rights are unacceptable. Still, at the present stage of human evolution, I think that we ought to stress obligations first. After we have defined our duties and responsibilities, we can examine and defend our rights and

166

entitlements. Let me, however, note in this connection that, for instance, the United Nations Human Rights Charter, although designed to serve the worthy purpose of protecting the weak and oppressed, may convey the erroneous impression that we human beings are the natural holders of rights without incumbencies, not even the incumbency of respecting other people's rights. In this respect, the usual formulation of the bills of rights seems unsatisfactory.

As human beings, we have many obligations to our fellows and to the other creatures, both living and unborn. This issue is easily forgotten and is raised only when we see civil rights patently trampled underfoot or animals grossly ill-treated or ruthlessly eliminated. I suggest that we must adopt a much broader view, particularly of *our liabilities toward the coming generations and future life generally*, bearing in mind that ignorance or disrespect of what should be incumbent on us today in our quality as well-informed, formidably-equipped and civilized people may well spell widespread disaster tomorrow.

IKEDA: In the past, it was necessary to work out bills of rights for the downtrodden common people who found themselves at the mercy of sometimes cruel and tyrannical rulers and privileged classes. I think I should emphasize for the sake of clarity that neither you nor I recommend giving up an inch of ground that has been won for the sake of ensuring those civil rights and liberties.

Now, in most parts of the world, the ordinary masses are in a position of privilege and power in relation to other living creatures, nature, and future generations. Therefore we must contain our own rights within a framework of obligation to prevent incursion on the rights of all these other parts of the world environmental system.

PECCEI: Let me reinforce the point by referring again to one of

167

the major responsibilities that has emerged in this new age: the reestablishment of good relationships with nature. The imperative of Earthcare and the careful husbandry of the world's resources is not merely a question of education, although we must certainly be educated to give it high priority. It is not enough, however, to recognize that we are wrong to destroy forests, overkill other species, pollute lakes, devastate environments and squander hydrocarbons. Our self-centered behavior has the effect of depriving posterity of the conveniences and comforts we seek for ourselves, as we impoverish our planet perhaps permanently or at least for a long time to come.

It does not suffice to realize that we must protect the capital of our natural wealth and learn to live on the income alone. If we are really to save the planet and our species, our commitment must be of a much higher order. We must face the reality that an additional couple of billion inhabitants will have to be accommodated on Earth by the beginning of the next century (followed by an even greater number thereafter); that the total world population will be even more unevenly distributed over the globe; that these people's demands for food, goods and services will soar much more rapidly than their numbers; and that antiquated concepts and practices of human supremacy, security, sovereignty, nationalism and tribalism will die hard, protracting perverse uses of our scientific knowledge and technological power.

It would be a tragic mistake to believe that such a world can host that population and offer it a modicum of well-being and dignity if each one of us does no more than refrain from certain actions. The conservation strategies that are currently being adopted in a growing number of countries are indispensable, and the stubborn resistance they encounter from all kinds of vested interests must be vanquished at all costs; but they are insufficient to keep our one and only support base – the Earth – in good shape in the face of growing human pressure.

IKEDA: Yes, and that pressure in the form of drains on resources and environmental pollution is making it hard for the Earth to support its present population and patently posing a tremendous threat to the existence of populations of the future. In spite of the knowledge that this is true, and in spite of warnings issued by people of knowledge and conscience like yourself, human beings continue their wasteful and destructive pursuit of present profit and comfort without taking into consideration the danger and hardship they are creating for their own children and grandchildren. Something must be done to alter this situation.

PECCEI: Unless a *radical, thorough reconceptualization of everything human and of man's place in his universe* occurs soon, the collapse of some natural systems is inevitable. To avoid this tragedy, we must define criteria to use and manage responsibly all the resources of the entire globe, and particularly its precious soils and fresh water, and to preserve its biological capacity from depletion or degradation. Large areas of land, vital ecosystems and adequate deposits of fossils and minerals should be kept untouched for this purpose, not only as a reserve for future requirements, but also as sanctuaries where nature can continue its evolution undisturbed, while other parts of the planet are actively exploited to satisfy current needs – such exploitation must always be carried out in a spirit of conservation and harmonious development of all peoples and nations.

All this, of course, is in the human interest and so it should be viewed. Yet it would be exceedingly naive to believe that the change can occur in time, unless we have a clear vision of how the human condition is declining all over the world, because of, *inter alia*, our divorce from nature. A new awareness of the menacing danger must shake us from our illusions and makes us ready to respond – in other words, we must ensure that the

169

human revolution arrives in time. This process can be facilitated if world public opinion and decision makers are brought to consider the enormous mass of problems that must be met and the need to approach them, not in isolation, but systematically by clusters in their regional and global contexts.

IKEDA: Yes, solving only one of the many complexly intertwined problems facing man today may generate new problems in other fields. In other words, all problems must, as you say, be approached in their own contexts. To give a simple example, reduction of fuel consumption suppresses the economy and gives rise to increased unemployment. What concrete steps do you recommend for the systematic approach to such problems in regional and global contexts?

PECCEI: At the Second World Congress on Land Use, held in June 1983 in Cambridge, Massachusetts, I proposed that a group of qualified nongovernmental organizations should prepare the outline for and study the feasibility of a stern *world plan and program of integral land and resource conservation, use and management*, in order to submit it to various international constituencies in the near future.

The rationale to be followed should be similar to that which has guided *l'amenagement du territoire* for the whole of France, but altered to the world scale and hence serving only to provide a very broad orientation. Moreover, the project should not be confined to the land masses but should embrace the oceanic expanse, which has been declared to be the Common Heritage of Mankind, and the polar regions as well. To start with, it should review and seek improvements to the new Law of the Seas that resulted from the mammoth UN conference on this subject and that, while opening the way for the establishment of an interim regime to regulate exploration, research and development on the deep seabed, left the door open to widespread pillage of marine resources, particularly by the stronger nations.

170

The project should also consider the possibility of making the Antarctic, a relatively virgin region, the touchstone of our commitment to keep some large parts of the planet in trust for all nations and peoples now and in the future. The 1961 Antarctic Treaty, which froze all claims on its resources for thirty years, expires in 1991. The time has come to insist that the negotiations, already started in the mood more of partition of spoils than of conservation, be conducted in the spirit of safeguarding this last free region of the world and fulfilling our obligations toward our children and children's children. To transmit to the next generations the whole of the Antartic and surrounding seas in a practically pristine state, without impoverishing us in any way, will be a token of respect for Mother Nature, the perennially renewing source of all life. Hopefully this will also encourage our successors to protect forever the southernmost part of the planet as the uncontested reign of wilderness, separate from the kingdom of man.

IKEDA: Whereas I certainly agree with you that the protection of Antarctica and the surrounding seas is important, I feel that it is equally important to create nature preserves in other, more readily accessible zones, where human beings can come into contact with the pristine world of nature, not in a resort with sophisticated equipment and comforts, but in its unaltered form. Some of the places I have in mind include certain zones in Japan – the Sanriku shore; Mount Daisetsu in Hokkaido; the mountains of central Honshu; and the coast of the Sea of Japan – as well as others in different parts of the globe – the Amazon Basin, the grassy plains of Africa, the Alps, and the islands of the Pacific.

PECCEI: The campaign of research and debates required to formulate the foundations of a planetary ecological strategy reconciling conservation and exploitation will have far-reaching moral, political and economic impact. By stressing our obligations toward the coming generations, and our own self-nterest in keeping the global environment healthy and productive, such a plan will elevate our consciousness not only of the new bounds of solidarity that must unite all humans and the responsibility they have in regard to life in general, but also of two other basic facts. The first is that no peace with nature will ever be possible unless peace reigns among men and, two, that Earthcare will foster, not stifle, worldwide economic activity and created a broad range of new job and self-employment opportunities for people otherwise condemned to inactivity.

Such a planetary project will compel us to recognize the changed realities and imperatives of our time, thereby helping us to purify our minds and hearts of viruses and toxins we have accumulated in the period of our blind exaltation of the material revolutions. Looking at our globe in its supreme beauty and generosity, we will understand that, to preserve it for our very existence and our joy and, at the same time, continue to benefit from whatever good the material revolutions can give us, we must make our desires subservient to higher designs respecting everything our Earth represents or needs. This can only be done by blending our desires with the human revolution.

7. More than Just Doing Things

IKEDA: The expenses of universal education inevitably demand the assistance of the state, but states often concern themselves only with training people proficient in technical or other practical fields. They want to make sure that the products of the educational systems they sponsor are politically and ideologically docile and subservient. Since I feel that this condition prevails in Japan today and lament it, I have insisted that the Soka Gakkai establish private university, high schools, middle schools and primary schools. Do you agree with me that states must not attempt to educate only skilled cogs for their own industrial or political machines and must try to cultivate well-rounded, self-reliant, individual human beings? What are your ideas on the optimum direction for education in the future?

PECCEI: I myself think of the world and its four and a half billion inhabitants in global terms and believe that the diversity of humankind forbids the uniquitous application of any one educational philosophy. What is good for Japan may not be equally good for Italy or Nigeria. Moreover, in many countries there are simply not enough private funds to start schools; therefore, state involvement in education is and will remain essential. Let me add, however, that there are many democratic nations in which the state school system is itself very democratic, as is the case, for instance, in my country, where criticism is sometimes raised at the partisanship and confessionalism of schools run by religious institutions.

Yet, whatever the educational system, the humanities are sacrificed nowadays for technical or scientific studies and for training in all fields of the economy, which is considered the main pillar sustaining modern societies. In general, contemporary schools teach how to do things. I for one am firmly against the priority and predominance given to disciplines dealing with the world of the physical, the measurable, the producible and consumable, since this implicitly means discriminating against the world of ideas. What we need most of all, in my view, is to learn how to think in terms of wholes, to reason about quality, to compare the imponderable, to ponder on the metaphysical, and to appraise and judge critically the ensemble of our knowledge, information and current behavior – in order to distill a sensible philosophy of life.

8. Education and Learning

IKEDA: Before leaving my topic, I should like to hear your opinions on education beyond the scope of ordinary schooling. Often the most idealistic home or school education becomes impotent when confronted with the sordid realities of society. I believe that social conditions must be altered so that young people can put into practice in their professional life the precepts they learn at school, in an atmosphere in which intellectual, artistic and spiritual fulfillment is an ideal worthy of respect. Leaders in all fields should be equipped spiritually as well as technically, and they should attempt to make the fullest possible use of their spiritual values in practical affairs.

To assist citizens who work to further their educational qualifications, adult education programs should be conducted by schools, libraries, museums and art museums. Scientists, artists, philosophers and religious leaders must devote part of their time to enlightening the public through lectures, discussions and seminars. Their efforts will be richly repaid by the impetus contacts with the general populace provide to their specialized research, thought and creativity. What are your opinions on this kind of adult education?

PECCEI: Your suggestions are excellent for education in developed countries where libraries, museums, schools and many other facilities are plentiful. They are not applicable, however, in vast regions of South-East Asia, the Indian subcontinent, Africa and parts of Latin America. In most of the Third World not only are there no means worth mentioning available for adult education, but also hundreds of millions of children never have the opportunity of enjoying formal schooling of any kind. And yet the overall military allocations are considerably higher than those destined for educational purposes.

174

We of The Club of Rome are of the opinion that it is important to make a distinction between what is conventionally termed *education* and *schooling* on the one hand, and what we mean by *learning* on the other. 'For us, learning means an approach, both to knowledge and to life, that emphasizes human initiative. It encompasses the acquisition and practice of new methodologies, new skills, new attitudes and new values necessary to live in a world of change. Learning is the process of preparing to deal with new situations. It may occur consciously, or often unconsciously, usually from experiencing real-life situations, although simulated or imagined situations can also induce learning. Practically every individual in the world, whether schooled or not, experiences the process of learning – and probably none of us at present is learning at the levels, intensities and speeds needed to cope with the complexities of modern life'.[1] Totally illiterate people can learn to be full-fledged human beings while conserving a great purity of spirit and having acquired a sound view of the world they must live in. At the other end of the spectrum, well-educated people sometimes never learn a thing about real life, and the most brilliant scholars may be tempted to use their knowledge to puruse perverse goals or suffocate the freedom of others.

I maintain that even the highest levels of education are less valuable than they might be, if they are not complemented by learning geared to the complex and changing realities of our total environment.

IKEDA: I agree with you, of course. As you say, the emphasis in modern education is on science and technology and I have already said that I feel much of what students learn in school is useless to them in later life. This is not as true of scientific and

[1] *No Limits to Learning, A Report to The Club of Rome*, by J. Botkin, M. Elmandjra and M. Malitza, Pergamon Press, Oxford, 1979. (In German: *Das Menschliche Dilemma*, Verlag Fritz Molden, Wien and Munich).

technological learning as it is of the humanities and of such things as morality and ethics – that is, the studies of what it means to be a human being and the obligations of the human condition. When university students receive any ethical or moral training at all, it is of the kind that tells them to live in an honest, upright way. Then, when they are on their own in society, they often find themselves in conditions in which honesty and uprightness lead to failure.

We must not allow ourselves to accept as inevitable a society in which this is true. Instead, we must try to alter that society; to enable us to do this, a new kind of learning is essential.

PECCEI: Until recent times, the traditional learning experience was good enough if it helped to buttress an existing system or maintain an established way of life. In our epoch of great transitions, this is no longer sufficient and the demands made on each of us have become more exigent. A new type of *innovative* learning is indispensable, for we must endow ourselves with new capacities that can bring the needed change into our societies, update our behavior and restructure our insitutions. Then we must more realistically assess the *problematique* we must meet and the best way of meeting it. Linear and phased approaches are no longer of any avail: we must acquire the habit of adopting a systemic approach, embracing simultaneously the condition and dynamics of large contexts. Moreoever, our capacity to learn can no longer draw its main inspiration from experience gathered in a past that differed vastly from the present and the likely future; instead it must be eminently future-oriented and, as far as possible, *anticipatory*. It must be *participatory* as well, because of the need for the involvement not only of decision-making elites, but also of the very large strata of the population in the endeavor to lift humankind to a higher threshold of awareness and responsibility.

Much research is going on, and several pilot projects have

176

been initiated in various parts of the world, to discover the mechanisms whereby people learn, to test the learnability of the average human being and to stimulate the capacities of understanding and creativity of such different groups of people as housewives, children, students, military recruits, state functionaries, and so on. In spite of the hope that all these efforts will cast light on the fundamental importance of a new learning to improve our condition, it goes without saying that learning can never be a substitute for education; it must be a complement to it and can be a catalyst in the regeneration of many benefits of education that are now being lost. Yet the function of learning must not be minimized. In a world confused about the relevance of what it is doing and what it can and should do, a sound foundation of common-sense learning can help individuals and societies to make a no-nonsense appraisal of the terrain they are treading and of what lies ahead.

9. No Decelerating Now

IKEDA: Today man is not a pilot flying an aircraft through clear, blue skies but a driver in an automobile hurtling down a winding road through a forest. As you have said, quick decisions combined with a long view of future consequences are vital in dealing with the modern situation. However changes occur today with such violence that the amount of future we can foresee is very small indeed. When prompt decisions are required, our low visibility can lead to disaster. Scientists and technicians are eagerly trying to accelerate the pace of change in our world. I think man, the driver on the winding road, must not accelerate but must decelerate – at least to a velocity enabling him to cope with unexpected curves. What is your opinion?

PECCEI: We *are* driving at a reckless speed on a winding road, risking catastrophe at any moment. Yet if speed stands for technology, we cannot decelerate – de-technologize – our onward rush. In reality retreat from a technological civilization to a simpler society is well-nigh impossible and complexity breeds more complexity. We are thus caught in a vicious circle, obliged to rely willy-nilly on bigger organizations, larger bureaucracies, more complicated mechanisms and greater automation – all requiring increasingly advanced technologies. At the same time, our problems of environment, energy, food production, urbanization, social justice, security, unemployment, crime, alienation, and so on continue to ramify into one another, enmeshing and interacting in a disorderly way to form what we have called the global *problematique*. To meet their challenge we have created a host of complex urban, industrial, information, communication, traffic, monetary, educational and military systems that, more often than not, compete with one another for money and resources or serve divergent goals and obey different logics. Moreover, under the pressure of growing demands, several of these systems become congested and give signs of overheating or possible disruption; experience tells us that, because of the domino effect, a breakdown in one of them may contaminate the others.

Therefore, at present, *we are propelled ahead by largely unbridled factors* and can hardly hope to be able to decelerate. What we can and must do is to control these factors better by changing gear and driving, not in a state of self-elation over our ability, but with a sober, responsible mode of behavior. We can and must try to do what we have not done so far (although this negligence is almost unbelievable): chart our way ahead or at least familiarize ourselves with the alternative routes that may be available, so as to prepare for our journey, and then proceed more safely.

IKEDA: I cannot agree that deceleration is out of the question.

You say that we must break out of our state of self-elation and strive to control the unbridled factors propelling us along the curving, perilous road. In my opinion, the moment we wake from that state of self-elation, we must slow down – even stop – take a look around to orient ourselves accurately, determine the optimum direction for future progress, and then go ahead at a sensible speed. You are correct in saying that many factors connected with our speed are intertwined in complex ways, but it is always mankind who has a foot on the accelerator. And man can, if he is in his right senses, reduce the pressure on that foot or even apply the brakes.

PECCEI: To continue this metaphor, even if we cannot as yet drive as reliably as we should and predict or select which course we will actually follow, we can, however, indicate with some degree of accuracy the itinerary we want to take and can probably pursue, depending on the circumstances and on the decisions we make at certain strategic points. We are not totally helpless.

We are in a position to project demographic growth curves with some assurance for the next couple of decades, extrapolate certain trends in other fields, and make several technological forecasts, all of which can give us an idea of the kind of situations we are likely to create in the not-too-distant future. We know that, if we continue to pump large quantities of carbon dioxide into the air through the exhausts of our automobiles and by burning coal, the temperature of the entire atmosphere will rise and that, if more powerful arms are stockpiled by more and more countries, these arms will some day be used whereas, if we reduce armaments, we can expect a more peaceful future. We know too that, if we carry out a modicum of planning to rationalize the use of raw materials, stabilize currencies and manage manpower, we can cushion ourselves somewhat against economic crises that will otherwise certainly hit us all the harder.

179

In other words, even if it is true that we are as yet unable clearly to state our goals and have not thought much about our future, *we have the possibility of guiding ourselves somewhat better than we are currently doing.* Guiding ourselves better in the general direction of predetermined goals is the objective of the *Forum Humanum* project sponsored by The Club of Rome, on which, as I have already mentioned, groups of young people are working. If some of the things described in the previous pages are accomplished, we will be in a position to map and control our course to a certain extent, while progressively equipping ourselves more adequately to live in an increasingly technologized society, and developing our own qualities and capacities so as to be able eventually to take our destiny in our own hands.

10. Not a Goal, but a New Course

IKEDA: Yes, and that means the human revolution. The human revolution is more a change of course than the achievement of a goal. Through it, we come to understand where the goal is and attempt to gain it, though perfection in this world is impossible. Although people devoted to the human revolution make perfectly ordinary mistakes, inside they are changed. Their distinctive qualities become apparent in the long run. I regard the human revolution as a continuing journey, not as the reaching of a destination. Do you agree?

PECCEI: I subscribe to your point of view. How much more stimulating it is to perceive that one is steering his or her own course and beginning to travel toward the right goals – even distant ones – than to reach them, even if this were possible. The truly amazing human venture, as long as it lasts, is probably destined to renew its goals continuously and to devise ever newer ways and means of attaining them, moving from one phase to a more advanced one with new goals and new possible solutions, and so on. What is important for our generations is not to remain on the present wrong course too long or to meander about blindly without worthy goals in mind, for we have within our powers the possibility of quick, sure, albeit involuntary self-debasement or self-destruction.

At present, we find ourselves at *one of the danger points of human history*. We have created a conjuncture more ominous than that when, nearly a millennium and a half ago, most of the civilized world was plunged into the Dark Ages. Adverse phenomena all over the world are at a peak never reached before. The same is true regarding the precariousness and instability of many situations, compared with the unparalleled exigencies and expectations of our time and the far larger number of people involved. The gigantic mass of problems we are unable to cope with, or even to understand, are likely to grow much bigger and become much more complex. Humankind is propelling itself along this catastrophe-bound course under the impetus of forces that have escaped its control or have been engendered in reaction to its actions.

Disaster of one kind or another can occur even in the near future, much earlier than anyone has ever feared. And it will occur ineluctably — beginning here or there and then becoming generalized — *unless we act immediately to change our course*. Acting at once is imperative, because with every year this supreme task will become more arduous. It will be much more difficult to take steps to ensure salvation in 1985 than it was in 1980. This decade is certainly crucial.

I think, in conclusion, that the severity of our plight can no longer be denied and that the warning we and others untiringly give about the extreme gravity of the global situation and trends and the absolute need to stop and reverse them must not be ignored. After making this premise unequivocably clear, I want

181

to reiterate with equally firm conviction that *it is within our power to turn the tide*. There can be no doubt that our generations, who are responsible for conducting human affairs at this historical juncture, possess all the knowledge and means to overcome the obstacles and negative circumstances that have made our situation difficult all over the world, and to reverse the downward drift of the human condition. Failure to recognize the growing grimness of the state of the planet would be irresponsible; failure to believe that we possess the objective possibility of improving that state or refusal to exert supreme effort to bring about such a change would be even worse. Both attitudes must be condemned as a grave mistake that would have unfathomable consequences for future human history.

All of us can contribute to avoiding these mistakes. And, as the human revolution is the key to positive action leading to the adoption of a new course and the revival of human fortunes, we must do whatever is in our power to help set it in motion — *before it is too late*.

CONCLUSION

Instead of comparing our individual presentations or extracting the gist of our dialogue to draw conclusions at the end of the book, we think it is fairer to allow the reader to evolve his or her own overall interpretation of the meaning and import of what we have said and to discuss and evaluate on the basis of his or her own criteria the ensemble of opinions, concepts, ideas, and suggestions we have expressed. It seems to us that this is the best way to encourage reflection and foster debate among the greatest number of interested people on at least some of the crucial issues we have touched upon. Since the book will appear almost simultaneously in several languages, we are mildly optimistic that an expansive and hopefully elucidating exchange of views will take place in diverse quarters. We are sure that we ourselves will benefit immensely if such multiple debates occur because any issue looks different and presents a variety of approach methods when examined from many angles.

Experience shows that world public opinion is maturing rapidly. Now that we understand that humanity as a whole faces fundamental alternatives which promise to alter our destiny completely, thus making the sharing of a common fate ineluctable, people everywhere are eager to participate in a broad-based debate on the human conditions and the prospects ahead. We therefore leave it to the reader to interpret how much more complex and disconcerting life is now than it was in the past and to decide what each person can or should do, in cooperation with his peers, to improve the common lot.

One issue, however, stands out and should unite us all in a supreme effort. The issue is peace, which, though deserving maximum attention and soul-searching analyses, lamentably too often receives neither. While proclaiming it as our cherished goal, in our daily rounds we do many things that put peace out of our reach. We seem to suffer from a kind of schizophrenia which produces an unconscious dichotomy between what we earnestly invoke, on the one hand, and what we actually provoke (or promote) on the other. Conquerors of the planet, possessed of more knowledge than we know how to employ, we are still too insecure to live in peace. As individuals, communities, nations, races and religions, we are harried and torn within and among ourselves. Furthermore, torment and strife characterize our relations as a species with most other forms of life and with nature at large.

In spite of being miserable and afraid because of the nonpeace we experience within ourselves and in our connections with almost everything around us, we seem unable to cease acting in ways that constantly aggravate our situation. Indeed, at no time in history has our globe been so gravely riven with endemic warfare, military and civil violence, widespread torture and terrorism, overt or incipient destruction and scientific preparation for still further havoc.

The situation is not, however, devoid of flickering rays of hope. Increasing numbers of men and women of all cultures and conditions are becoming aware of the monstrous, rash folly that has brought us to this pass. The yearning for saner human behavior and the desire for a humane life quality that have already begun appearing everywhere are sure signs that humankind is waking to the danger and can, given the will to do so, reverse the current downward trend.

Instead of requesting you to share our views on most subjects, we invite you to criticize and improve them. However, on the

184

issue of peace, we demand that you join us, close ranks with us, and do your utmost to help redress a situation that is rapidly becoming desperate.

With different environments and capacities, we two authors nonetheless devote the best of our energies to studying ways and means for global pacification. No matter how you may disagree with us on many other points if, after having read these pages, you want to join us in the struggle for peace, we could hope for no better reward for our labors.

Index

186

187

Europe, 11, 80, 84, 95, 106, 108, 109, 16i1 forests in, 42
European Community, see EEC
Euthropication, 65

Faith, 150
Falklands-Malvinas war, 100
Fauna, 17, 23, 432-3, 432-3, 47, 49, 50, 51, 63, 67; Buddhist view of, 67-8; spirituality of, 66
Federalism, 101, 108, 164
Feudalism, 106
Fifth generation computers, 158
Flora, 17, 23, 42-3, 47, 49, 50, 51, 63
Folly, 119, 120, 149
Food, 54, 59, 60, 61, 62, 102, 168; population growth, 56-7, 58
Forum Humanum, 147-8, 180
Fossils, 169
France, 80, 81, 96, 102, 112, 170
Freedom, 93-4
Free will, 91-2, 93
French Revolution, 86
Fuel, 21, 22, 31, 35, 43

Genetic code, 28
Genetic engineering, 29, 31, 153
Geothermal energy, 40
Global consciousness, 137, 139, 170
Global government, 99, 100-1
Globality, 12, 13
Goals, 180-2
Gold miners, 43
Good and evil, 87, 88, 111, 127, 138
Grassroots movements, 146
Greece, 51-2, 81
Greed, 17, 23, 49, 51, 119, 120, 149, 159
Growth, 29, 33, 34, 35

Hatred, 119, 120
Hell, 127
Himalayas, 65, 78-9
Hiroshima, 157

Hispanic Americans, 115
Historical time, 15
Hitler, Adolf, 91, 93
Hominization, 14
Homo sapiens, 14-15
Honshu Mountains, 171
Horn of Africa, 100
Human nature, 24-5, 85-6
Human potential, 140-1, 142, 144, 149, 152-3, 155, 157-8
Human relationships, 73
Human revolution, see Inner revolution
Human rights, 145-6
Hunger, 122
Hunting, 23, 51, 52, 53
Huxley, Julian, 160
Hypnosis, 79-80

Ibis, 53
IBM, 107
Ignorance, 140
Immigration, 113
India, 54, 65, 82, 94, 112, 164, 174
Indian Buddhism, 87
Industrialisation, 42, 58-9, 60-1
Industrial revolution, 16, 27-8
Inequality, 74
Inner disorder, 134, 135, 140
Inner revolution, 85-6, 105, 119, 124, 125, 128, 129, 135, 136, 140-1, 144-5, 146, 147-8, 149, 169, 170, 180-2
Instincts, 76-7, 78
Intelligence, 158-9
Intercontinental warfare, 29
International economic order, 133
Internationalism, 109-110
International Youth Year, 148
Iraq-Iran war, 100
Islam, 117, 119
Israel, 100
Italy, 47, 48, 52, 108, 162, 164
Ivory, 50, 51

189

Wind energy, 40
Wisdom, 124, 134
World *problematique*, 12, 13, 29, 137, 176, 178

World War I, 84
World War II, 74, 83, 84, 97, 99, 103, 124, 154
Wrath, 149